Mathematical Basis of Statistics

This is a volume in
PROBABILITY AND MATHEMATICAL STATISTICS

A Series of Monographs and Textbooks

Editors: Z. W. Birnbaum and E. Lukacs

A complete list of titles in this series appears at the end of this volume.

Mathematical Basis of Statistics

JEAN-RENÉ BARRA
UNIVERSITÉ SCIENTIFIQUE ET MÉDICALE DE GRENOBLE

Translation edited by

LEON HERBACH
POLYTECHNIC INSTITUTE OF NEW YORK

1981

ACADEMIC PRESS
A Subsidiary of Harcourt Brace Jovanovich, Publishers
New York London Toronto Sydney San Francisco

TO ELIETTE AND MILDRED

ACADEMIC PRESS, INC.
111 Fifth Avenue, New York, New York 10003

United Kingdom Edition published by
ACADEMIC PRESS, INC. (LONDON) LTD.
24/28 Oval Road, London NW1 7DX

Library of Congress Cataloging in Publication Data

Barra, Jean-René, Date
Mathematical basis of statistics.

(Probability and mathematical statistics)
Bibliography: p.
Includes index.
1. Mathematical statistics. I. Herbach, Leon H.
II. Title.
QA276.B2878 519.5 80-519
ISBN 0-12-079240-0

AMS (MOS) 1970 Subject Classifications
Primary: 62-02
Secondary: 60E05, 60E10, 62A20, 62A25, 46-01

PRINTED IN THE UNITED STATES OF AMERICA

81 82 83 84 9 8 7 6 5 4 3 2 1

Contents

Chapter 3 Statistical Information

Chapter 4 Statistical Inference

Chapter 5 Testing Statistical Hypotheses

Chapter 6 Statistical Estimation

Chapter 7 The Multivariate Normal Distribution

Chapter 8 **Random Matrices**

Chapter 9 **Linear-Normal Statistical Spaces**

Chapter 10 **Exponential Statistical Spaces**

Chapter 11 **Testing Hypotheses on Exponential Statistical Spaces**

Chapter 12 **Functional Analysis and Mathematical Statistics**

Appendix **Conditional Probability**

Foreword

Because mathematical statistics is of great interest—practical as well as theoretical—many statistical books have been published, as indicated in the reference list of this book. Most of these books are excellent, but they generally aim at making application easier; the systematic analysis of statistical spaces is missing.

The present work by Professor J.-R. Barra [*Notions Fondamentales de Statistique Mathématique*, Dunod, Paris, 1971] is devoted to the systematic analysis of fundamental statistical spaces, from the point of view of modern mathematics. In this manner, the author analyzes these spaces in great detail and depth, first without being linked to decision problems, and then linking them to decisions and strategies. Thus, for example, we find here, for the first time in the literature, the formal definition of nuisance parameters with respect to a statistical decision problem.

The chapter on functional analysis and mathematical statistics is beautiful . . . and J.-R. Barra's work is an excellent mathematical introduction to modern mathematical statistics. It will be very useful to students who desire to be introduced to statistics by a thorough study of the most important statistical spaces, in order to extend their studies to sequential analysis and other special branches of modern mathematical statistics.

YU. V. LINNIK

Editor's Preface

The book contains gallicisms, some because the Bourbaki school's notation has proved useful in many branches of mathematics and should be used more frequently in statistics. Several examples follow. Inverse brackets are used, so that the half-open interval is indicated by $]a,b]$, rather than $(a,b]$. The gradient is indicated by $\operatorname{grad} f$, rather than the more common ∇f. The expression *à valeurs dans* $(\mathscr{A}, \mathscr{B})$ was originally translated as *valued on* $(\mathscr{A}, \mathscr{B})$. However, it was felt that this might sound strange to an English-speaking group and, in most cases, has been changed to *having* or *taking values on* (or *in*) $(\mathscr{A}, \mathscr{B})$. A better translation might be *with range* $(\mathscr{A}, \mathscr{B})$. Topologists might object to confounding of *in*, *into* (or *on*, *onto*), which sometimes occurs. The distinction is absent in French, where both concepts are expressed by *dans* (or *sur*). The English-speaking mathematician who feels superior because this distinction is possible only in his language should be reminded that, while no distinction is made in English between the two *sufficient's* as in *sufficient* statistic and necessary and *sufficient* condition, the French statistician would use *statistique exhaustive* and *condition nécessaire et suffisante*.

Other changes in terminology are intentional. For example, a *similar* test is called (for example, in Section 5.1) a *free* test. This is because the concept of free test is related to the concept of mean-freedom of a statistic (Section 2.2); Barra believes that the terminology should be similar (no pun intended).

Some variations from American usage are noted in footnotes. Minor errors in the French edition have been corrected. Of course, the editor takes the blame for any new errors introduced. The editor profited from discussions on the original manuscript with his colleagues, Arthur Greenwood and George Bachman. In addition, he wishes to thank the staff of Academic Press and the typesetter who did admirable jobs with an imperfect manuscript.

Lisbon L. H.

Preface

This book is intended for graduate and postgraduate students. However, it is necessary to be familiar with the fundamental and elementary techniques of modern probability theory. To help the reader, we have collected in an appendix those conditional probability results that will be used frequently in the book.

The purpose of the book is to enable the reader to become acquainted with the methods and the mathematical basis of statistics. For this reason, we have limited ourselves to the procedures and fundamental notions expressed in a modern formalism. We have thus used the language of modern probability theory. The reader will note that the notion of statistical space is fundamental to this formalism. Already introduced by other authors (Le Cam, for instance), the systematic use of this notion allows a more concise and precise formulation of problems. From a theoretical point of view, the notion permits one to make useful comparisons with other mathematical theories. On the other hand, from a practical view, the concept also allows one to better and more clearly express the conditions and hypotheses related to a statistical experiment.

The first three chapters deal with the notion of statistical space and general, related concepts. They are useful for the presentation and mathematical study of statistical data. All fundamental notions are therefore introduced here provided that they are not related to decisions to be taken when these data are known. The general scheme of a statistical decision is given in Chapters 4–6, with the study of the two most common types, estimation theory and nonsequential tests of hypotheses. In Chapters 7 and 8, we develop the fundamental techniques of probability theory connected with normal distributions on finite-dimensional linear spaces—the ones most used in applications. Chapters 9–11 are devoted to the study of the most important types of statistical spaces—linear-normal and exponential. Most applications of mathematical statistics are special cases of general results obtained in these chapters. The last chapter, which is very theoretical, gives some examples of applications of functional analysis in statistics. The chapter on numerical problems in statistics, including some ALGOL programs, which was in the French edition, has been omitted since such computer programs are now commonplace.

Equations, theorems, etc., are numbered anew in each section. When necessary, the section, and sometimes the chapter also, is appended to the number. For example, Theorem 2 in Section 3 of Chapter 7 is referred to as Theorem 2 in that section, as Theorem 3.2 in another section of the same chapter, and as Theorem 7.3.2 in any other chapter. References to the Appendix are prefixed A.

The reader is strongly encouraged to do the exercises in order to absorb the material in the corresponding chapter. Similarly, the proofs of theorems are usually very concise and the reader should take the time to reread them, filling in the gaps as additional exercises. The reader who would like to improve his knowledge of probability theory is advised to begin this book with Chapter 7 and its exercises.

Finally, I wish to acknowledge again all those who have contributed to the success of the French edition, and also to thank Prof. E. Lukacs and Academic Press, who accepted the manuscript, and my friends, Leon and Mildred Herbach, whose help was invaluable in realizing this adaptation.

Grenoble J.-R. B.

Notation and Terminology

The numbers preceding symbols indicate the chapter and section in which the notation is introduced. Exercises are preceded by an E.

E1.5	$\mathscr{B}_{\mathscr{X}}$	σ-field of the Borel sets of the finite-dimensional space $\mathscr{X}$
E1.10	$\mathbb{R}^+$	space of nonnegative real numbers
E1.10	$\mathscr{B}_{\mathbb{R}^+}$	σ-field of the Borel sets of $\mathbb{R}^+$
7.2	$\mathbb{1}_n$	unit matrix of order n
1.3	1_E	indicator function of the set E
9.1	$\mathbf{1}_{\mathscr{X}}$	unit quadratic form of a Hilbert space $\mathscr{X}$
E1.5	$N(m, \Lambda)$	normal distribution having mean m and covariance matrix Λ
E6.7	$N(z)$	cumulative distribution function of the $N(0, 1)$ distribution on $\mathbb{R}$
1.3	$m \otimes m'$	product of the measures m and m'
8.1	$A \otimes B$	Kronecker or direct product of matrices A and B
1.3	$\mathscr{A} \otimes \mathscr{B}$	σ-field generated by sets $A \times B$, where $A \in \mathscr{A}$ and $B \in \mathscr{B}$ ($\mathscr{A}$ and $\mathscr{B}$ are two σ-fields)
2.5	$A \triangle B$	"exclusive union" of sets A and B, $A \triangle B = A \cup B - A \cap B$
7.1	$P * P'$	convolution of two distributions P and P'
8.1	$\mathscr{S}_n$	space of symmetric matrices of order n

8.3	$\mathscr{S}_n^+$	space of symmetric matrices of order n which are nonnegative definite
E1.10	$\mathscr{P}(\lambda)$	Poisson distribution having parameter λ
E2.2	$\mathscr{B}(n, p, q)$	binomial distribution having generating function $(q + pu)^n$
E1.10	$\Gamma(x)$	gamma function defined for $x > 0$ by $\int_0^\infty e^{-t} t^{x-1}\, dt$
1.2	P_X	distribution of the random element X
2.3	${}^tM$	transpose of the matrix M
2.3	Λ_X	covariance matrix of the random vector X
4.5	$\inf(f(x) \mid x \in E)$	greatest lower bound of f on E
4.5	$\sup(f(x) \mid x \in E)$	least upper bound of f on E
2.3	$\langle x \mid y \rangle$	inner product of two elements of a Hilbert space
8.1	$\langle A \mid B \rangle$	inner product of matrices A and B
7.2	*order of a matrix*	the ordered pair defined by its number of rows and its number of columns (in the case of a square matrix we indicate only the number of rows)
1.2	*vector*	an element of $\mathbb{R}^n$, which we associate with a column matrix of order $(n, 1)$
1.2	*random variable*	a random element taking values on $(\mathbb{R}, \mathscr{B}_{\mathbb{R}})$ (i.e., the infinite values are excluded)
1.2	*random vector*	a random element taking values on $(\mathbb{R}^n, \mathscr{B}_{\mathbb{R}^n})$. For a random vector, we distinguish between its covariance, which is a quadratic form, and its covariance matrix, which is the matrix corresponding to the covariance with respect to a given basis

In some cases we describe a function f by the corresponding mapping

$$x \to f(x).$$

We may write P-a.e. for almost everywhere with respect to the distribution P, and we say that two random variables are P-equivalent if they are P-a.e. equal.

CHAPTER 1

Statistical Spaces

The notion of statistical space in mathematical statistics is an extension of the concept of probability space in probability theory, in which the probability measure is replaced by a *family* of probability distributions. The same problems arise in choosing a fundamental statistical space as in choosing a fundamental probability space in probability theory. Furthermore, in this chapter we note other analogies with probability theory: statistic and random vector, the products of statistical spaces and of probability spaces, etc.

1. Statistical Spaces; the Dominated Case

Definition 1 Let $\mathscr{P}$ be a family of probability distributions on a measurable space $(\Omega, \mathscr{A})$; $(\Omega, \mathscr{A}, \mathscr{P})$ is said to be a *statistical space*.†

† In Definition 1, the French edition used the notation $(\Omega, \mathfrak{A}, \mathscr{P})$, so that three alphabets were used: Greek for the sample space, German for the σ-field of subsets of sample space, and script for the family of probability distributions. Frequently, this convention was overridden and two of the spaces were symbolized using the same alphabet. The editor should take the brunt of any criticism for eliminating the German alphabet from the present text. It was done for typographic convenience and to avoid having to go through the text to remove the inconsistency referred to above. Since the components always occur in the same order, the reader should have no difficulty in distinguishing between the σ-field and the probability family. In addition, whenever the description of the family of probability distributions involves commas, either the family is enclosed by curly brackets or the comma before the last element of the statistical space is changed to a semicolon. (Ed.)

The space Ω, usually called sample space, is the set of observations. We assume that these observations are made at random with a probability distribution belonging to a known family $\mathscr{P}$. This fundamental assumption on the randomness of observations and the family of their distributions must be emphasized in any real statistical problem.

The family $\mathscr{P}$ is often described with the help of an index θ, called the *parameter*; in this case we write

$$\mathscr{P} = \{P_\theta, \theta \in \Theta\}.$$

Of course, to determine a suitable statistical space for any problem in mathematical statistics it will be necessary to examine in detail the experiment giving rise to that problem. As examples see Exercises 1–3.

Definition 2 The statistical space $(\Omega, \mathscr{A}, \mathscr{P})$, or briefly, the family $\mathscr{P}$, is said to be *dominated* if there exists a σ-finite and positive measure μ on $(\Omega, \mathscr{A})$ such that one of the following equivalent conditions is fulfilled:

(1) Every distribution belonging to $\mathscr{P}$ is μ-continuous.
(2) Every distribution belonging to $\mathscr{P}$ has a density of probability with respect to μ.

Theorems A.1.2 and A.1.4 establish the equivalence of these two conditions. We recall that a positive measure μ is σ-finite if Ω is a countable union of measurable events having finite μ-measure.

If for a dominated statistical space we choose a parameter θ and for every θ a determination $p_\theta(\omega)$ of the density $dP_\theta/d\mu$, we use the notation

$$(\Omega, \mathscr{A}; p_\theta, \theta \in \Theta)$$

and define the real valued function $\mathscr{L}(\omega, \theta)$, defined on $\Omega \times \Theta$, to be the *likelihood function*, where†

$$\mathscr{L}(\theta, \omega) = p_\theta(\omega).$$

The measure μ which dominates a dominated statistical space is not uniquely defined; if μ is absolutely continuous with respect to μ', μ' is also such a measure and the density is given (Section A.1) by

$$\frac{dP_\theta}{d\mu'} = \frac{dP_\theta}{d\mu}\frac{d\mu}{d\mu'}$$

The following theorem shows that in the dominated case we can always

† The reader will note that this notation is different from Fisher's $\mathscr{L}(\theta, \omega)$. Here the likelihood is treated as a function of *two* variables and not, as in Fisher, as a function of the variable θ, depending on ω. Sometimes the likelihood will appear as $\mathscr{L}(\omega, \theta)$ defined on $\Theta \times \Omega$. The order of the variables is unimportant and the two versions are used interchangeably.

choose, as measure μ, a probability distribution having interesting properties. This will be useful for theoretical computations.

Theorem 1 *A statistical space $(\Omega, \mathscr{A}, \mathscr{P})$ is dominated if and only if there exists a probability distribution P^* on $(\Omega, \mathscr{A})$, which dominates $(\Omega, \mathscr{A}, \mathscr{P})$, such that*

(a) *P^* is absolutely continuous with respect to every measure which dominates $(\Omega, \mathscr{A}, \mathscr{P})$;*

(b) *P^* is a strictly convex combination of a countable subfamily $\mathscr{P}'$ of $\mathscr{P}$; i.e.,*

$$P^* = \sum_{P \in \mathscr{P}'} c_P P \qquad \left(c_P > 0, \sum_{P \in \mathscr{P}'} c_P = 1\right);$$

(c) *P^* is equivalent to $\mathscr{P}$; i.e.,*

$$\forall A \in \mathscr{A}, \qquad [P(A) = 0, \forall P \in \mathscr{P}] \iff P^*(A) = 0.$$

We shall call the probability distribution P^* defined by the theorem a *special* probability distribution which dominates the structure. The proof of this theorem is based on the following lemma.

Lemma *The statistical space $(\Omega, \mathscr{A}, \mathscr{P})$ is dominated if and only if there exists a countable subfamily $\mathscr{P}'$ of $\mathscr{P}$ such that*

$$\forall A \in \mathscr{A}, \qquad [P(A) = 0, \forall P \in \mathscr{P}'] \Rightarrow [P(A) = 0, \forall P \in \mathscr{P}].$$

In this case, let P^* be a *strictly* convex combination of distributions belonging to $\mathscr{P}'$. Then P^* is a distribution on $(\Omega, \mathscr{A})$. Furthermore, if μ dominates $(\Omega, \mathscr{A}, \mathscr{P})$ we have

$$\forall A \in \mathscr{A}, \quad P^*(A) = 0 \iff [P(A) = 0, \forall P \in \mathscr{P}'] \iff$$
$$[P(A) = 0, \forall P \in \mathscr{P}] \Leftarrow \mu(A) = 0$$

and P^* satisfies the conditions of Theorem 1.

Corollary *If Ω or $\mathscr{P}$ is countable, the statistical space $(\Omega, \mathscr{A}, \mathscr{P})$ is dominated. If the family $\mathscr{P}$ is dominated, the family of distributions which are finite convex combinations of distributions belonging to $\mathscr{P}$ is also dominated. Finally, if $\mathscr{P}$ is a countable sequence of dominated families, the union of these families is dominated.*

This corollary is a straightforward consequence of the previous propositions. When Ω is countable, we call the σ-finite measure, whose value in every subset of Ω is the number of elements in this subset, the *counting measure*. In the usual practical cases the statistical spaces that occur are dominated either by a counting measure or in the continuous case by Lebesgue measure.

2. Statistics, Integrable Statistics, and Completeness

Definition 1 We define a *statistic* on the statistical space $(\Omega, \mathscr{A}, \mathscr{P})$ as a measurable mapping T from $(\Omega, \mathscr{A})$ to a measurable space $(\mathscr{X}, \mathscr{C})$.

It is very important to note that T is defined without including $\mathscr{P}$, that is, it must not depend on the parameter. If $\mathscr{X} = \mathbb{R}$, we say that T is a real valued statistic. If $\mathscr{X} = \mathbb{R}^n$, we say that T is an n-dimensional vector valued statistic. To the concept of a statistic in mathematical statistics there corresponds the concept of a "random element" in probability theory.

In practice, a statistic is used to extract some significant values from the initial data. For example, see Exercise 1 and the statistic defined by the frequencies r-tuple.

For every distribution P belonging to $\mathscr{P}$, the statistic T considered as a mapping from $(\Omega, \mathscr{A}, P)$ to $(\mathscr{X}, \mathscr{C})$ defines a random element and induces a distribution P_T. We write

$$\mathscr{P}_T = \{P_T, P \in \mathscr{P}\}$$

and call $(\mathscr{X}, \mathscr{C}, \mathscr{P}_T)$ the statistical space induced by T. Finally we write

$$(\Omega, \mathscr{A}, \mathscr{P}) \xrightarrow{T} (\mathscr{X}, \mathscr{C}, \mathscr{P}_T).$$

Definition 2 Let T_1 and T_2 be two statistics defined on $(\Omega, \mathscr{A}, \mathscr{P})$ and taking values in $(\mathscr{X}_1, \mathscr{C}_1)$ and $(\mathscr{X}_2, \mathscr{C}_2)$, respectively. We say that T_1 is *equivalent to* T_2, and write $T_1 \sim T_2$, if

$$T_1^{-1}(\mathscr{C}_1) = T_2^{-1}(\mathscr{C}_2).$$

Note that this definition does not refer to or depend on $\mathscr{P}$. Specifically if T_1 and T_2 are deduced from each other by a measurable one-to-one mapping, then T_1 and T_2 are equivalent.

Definition 3 We say that the event $A \in \mathscr{A}$, has *$\mathscr{P}$-measure zero* if and only if

$$\forall P \in \mathscr{P}, \qquad P(A) = 0.$$

The frequently used notation "$\mathscr{P}$-almost everywhere" ($\mathscr{P}$-a.e.), will be used as a short form for "except on a set of $\mathscr{P}$-measure zero."

Definition 4 Let T_1 and T_2 be two statistics defined on $(\Omega, \mathscr{A}, \mathscr{P})$ and taking values in $(\mathscr{X}, \mathscr{C})$. We say that T_1 is *$\mathscr{P}$-equivalent to* T_2 and write $T_1 \sim T_2$, if the event $\{T_1 \neq T_2\}$ has $\mathscr{P}$-measure zero.

Remark If the statistical space $(\Omega, \mathscr{A}, \mathscr{P})$ is dominated by a special distribution P^*, the following argument proves that an event has $\mathscr{P}$-measure zero, if and only if it has a P^*-measure zero. If A has a P^*-measure zero, it has a P-measure zero for every distribution P belonging to $\mathscr{P}$ because P^* dominates $\mathscr{P}$. Conversely, if A has a $\mathscr{P}$-measure zero it also has a P^*-measure zero, as P^* is a strictly convex combination of a countable subfamily of $\mathscr{P}$.

Definition 5 Two statistics X and Y defined on $(\Omega, \mathscr{A}, \mathscr{P})$ are *independent* if, for every distribution P belonging to $\mathscr{P}$, the random elements X and Y are independent.

As an example see Exercise 5.

Definition 6 The real-valued statistic X defined on $(\Omega, \mathscr{A}, \mathscr{P})$ is *integrable* if for every distribution P belonging to $\mathscr{P}$ the random variable X is integrable; i.e., X has a mathematical expectation which we write as $E_P(X)$.

If X is an n-dimensional statistic, X is called integrable if and only if every component of X is integrable.

Definition 7 The real-valued statistic X defined on $(\Omega, \mathscr{A}, \mathscr{P})$ is *mean-free* (or has zero expectation) if $E_P(X)$ exists and does not depend on $P \in \mathscr{P}$ (or $E_P(X)$ is zero for every distribution belonging to $\mathscr{P}$).

As an example see Exercise 6. We write, when there is no risk of ambiguity, $E(X)$ for the common value of $E_P(X)$.

Definition 8 We call β_X the *image* of the integrable statistic X defined on the statistical space $[\Omega, \mathscr{A}, \{P_\theta, \theta \in \Theta\}]$, where the function β_X is the function on Θ defined by

$$\beta_X(\theta) = E_{P_\theta}(X) = \int_\Omega X \, dP_\theta, \qquad \theta \in \Theta.$$

Definition 9 Let $(\Omega, \mathscr{A}, \mathscr{P})$ be a statistical space and let $\mathscr{B}$ be a subfield of $\mathscr{A}$. We say that $(\Omega, \mathscr{B}, \mathscr{P})$ is a *complete* (or *quasi-complete*) statistical space if every $\mathscr{B}$-measurable statistic having zero expectation (or if every $\mathscr{B}$-measurable and bounded statistic having zero expectation) is $\mathscr{P}$-equivalent to zero.

When there is no possibility of confusion, we say briefly that $\mathscr{B}$ is complete or quasi-complete instead of saying that $(\Omega, \mathscr{B}, \mathscr{P})$ is complete or quasi-complete. Obviously, if a statistical space is complete, it is also quasi-complete. However, we can have (see Exercise 8) a quasi-complete statistical space which is not complete.

Remark Let $(\Omega, \mathscr{A}; P_\theta, \theta \in \Theta)$ be a statistical space dominated by a special distribution P^*. This statistical space is complete if and only if

$$\int_\Omega X p_\theta \, dP^* = 0, \qquad \forall \theta \in \Theta \quad \Rightarrow \quad X = 0 \qquad P^*\text{-a.e.}$$

In particular, we shall establish (Section 10.2) a theorem concerning completeness of exponential statistical spaces.

3. Prior and Posterior Distributions

Throughout this section, we assume that a σ-field $\mathscr{F}$ is defined on Θ and that P_θ is a *transition* probability function on $\Theta \times \mathscr{A}$ (see Section A.6). This assumption is weak and is satisfied in most cases. We write

$$[\Omega, \mathscr{A}, \{P_\theta, \theta \in (\Theta, \mathscr{F})\}]$$

for a statistical space satisfying this assumption. The following lemma is easily deduced from Theorem A.6.2.

Lemma *If the likelihood function $\mathscr{L}(\omega, \theta)$ exists and is measurable as a function on $(\Omega \times \Theta, \mathscr{A} \otimes \mathscr{F})$, then P_θ is a transition probability function on $\Theta \times \mathscr{A}$.*

Theorem 1 *The image β_X of every integrable statistic X defined on the statistical space $[\Omega, \mathscr{A}; \{P_\theta, \theta \in (\Theta, \mathscr{F})\}]$ is an $\mathscr{F}$-measurable function.*

Proof If X is the indicator of the event A, $\beta_X = P_\theta(A)$ is $\mathscr{F}$-measurable. A linear combination of $\mathscr{F}$-measurable functions is $\mathscr{F}$-measurable; the limit of an increasing sequence of $\mathscr{F}$-measurable functions is $\mathscr{F}$-measurable. Applying these two operations, repeatedly, if necessary, shows that the image of every positive and integrable statistic is $\mathscr{F}$-measurable. Finally, an arbitrary integrable statistic is the difference of two positive integrable statistics. ∎

Remark Using Definitions 12.3.1 and 12.3.2 we can consider the mapping $\beta: X \to \beta_X$ as a linear mapping from $L_1(\Omega, \mathscr{A}, \mathscr{P})$ into $F(\Theta, \mathscr{F})$, which is the space of all real-valued and measurable functions defined on $(\Theta, \mathscr{F})$. We can also consider β as a mapping from $\Lambda_1(\Omega, \mathscr{A}, \mathscr{P})$ or from $\Lambda_\infty(\Omega, \mathscr{A}, \mathscr{P})$ into $\mathscr{L}_\infty(\Theta, \mathscr{F})$. Thus, it appears that the statistical space is complete (or quasi-complete) if and only if β is a one-to-one mapping from L_1 (or Λ_∞) into F (or $\mathscr{L}_\infty$). For some statistical spaces, β is a well-known analytic transformation (Taylor expansion, Laplace transform, etc). In these cases, one can use the well-known properties of these transformations to establish the completeness of the statistical space. For example see Exercise 10.

Definition 1 Consider the statistical space $[\Omega, \mathscr{A}, \{P_\theta, \theta \in (\Theta, \mathscr{F})\}]$. We call every distribution Q on $(\Theta, \mathscr{F})$ a *prior distribution* (or distribution *a priori*).

The Bayesian point of view in statistics is based on a prior distribution. This distribution represents the information on the parameter θ before the statistical experiment. In some cases this information is justified, but in others it can appear artificial. The controversy between the Bayesian and the non-Bayesian viewpoint in statistics is well known, but we do not consider it a mathematical problem. The reader interested in the Bayesian procedure is referred to Good [26] and its extensive bibliography.

Theorem 2 *For the statistical space of Definition* 1, *the mapping* β_Q^* *defined by*

$$\forall A \in \mathscr{A}, \qquad \beta_Q^*(A) = \int_\Theta P_\theta(A)\, dQ(\theta) \tag{1}$$

is a probability distribution on $(\Omega, \mathscr{A})$. *If the statistical space is dominated by the measure* μ, *the distribution* β_Q^* *is absolutely continuous with respect to* μ.

Proof Theorem A.6.1 shows that there exists only one distribution π defined on $(\Theta \times \Omega, \mathscr{F} \otimes \mathscr{A})$ such that

$$\forall T \in \mathscr{F}, \quad \forall A \in \mathscr{A}, \qquad \pi(T \times A) = \int_T P_\theta(A)\, dQ(\theta)$$

and β_Q^* is the marginal distribution on $(\Omega, \mathscr{A})$ associated with π. Furthermore, if the statistical space is dominated by μ we have

$$\{A \in \mathscr{A}, \mu(A) = 0\} \quad \Rightarrow \quad \{P_\theta(A) = 0, \forall \theta \in \Theta\} \quad \Rightarrow \quad \beta_Q^*(A) = 0. \qquad \blacksquare$$

Note that the relation

$$d\beta_Q^*/d\mu = \int_\Theta p_\theta(\omega)\, dQ(\theta),$$

is not necessarily true. However it is valid if $p_\theta(\omega)$ is a measurable function on $(\Omega \times \Theta, \mathscr{A} \otimes \mathscr{F}, \mu \otimes Q)$ (see Exercise 11).

Definition 2 If a transition probability function $\tilde{P}_\omega(T)$ exists on $\Omega \times \mathscr{F}$ such that both systems $(\tilde{P}_\omega(T), \beta_Q^*)$ and $(P_\theta(A), Q)$ generate the same distribution π on $(\Omega \times \Theta, \mathscr{A} \otimes \mathscr{F})$, then for every $\omega \in \Omega$ we call the distribution $\tilde{P}_\omega$ on $(\Theta, \mathscr{F})$ a *posterior distribution* (or distribution *a posteriori*).

Note that this last Bayesian concept, (posterior distribution) does not admit the same mathematical representation as the first Bayesian concept (prior distribution). It seems natural to refer to the transition probability function $\tilde{P}_\omega(T)$ as the "inverse" of the transition probability function $P_\theta(A)$ because we have

$$\forall A \in \mathscr{A}, \quad \forall T \in \mathscr{F}, \qquad \int_A \tilde{P}_\omega(T)\, d\beta_Q^* = \int_T P_\theta(A)\, dQ(\theta).$$

If the statistical space is dominated, if Ω and Θ are Euclidean and if we have chosen a determination of $p_\theta(\omega)$ which is $\mathscr{A} \otimes \mathscr{F}$-measurable as a function on $\Omega \times \Theta$, then Theorem A.7.1 shows that

$$d\tilde{P}_\omega/dQ = \left\{ p_\theta(\omega) \middle/ \int_\Theta p_\theta(\omega)\, dQ \right\} \qquad \mu\text{-a.e.}$$

and we obtain the Bayes formula.

4. Products of Statistical Spaces

Definition 1 Let $(\Omega, \mathscr{A}, \mathscr{P})$ and $(\Omega', \mathscr{A}', \mathscr{P}')$ be two statistical spaces. We call the statistical space $(\Omega \times \Omega', \mathscr{A} \otimes \mathscr{A}', \mathscr{P} \otimes \mathscr{P}')$, where

$$\mathscr{P} \otimes \mathscr{P}' = \{P \otimes P';\ P \in \mathscr{P},\ P' \in \mathscr{P}'\},$$

the *product* of the two given statistical spaces. We write this product as $(\Omega, \mathscr{A}, \mathscr{P}) \otimes (\Omega', \mathscr{A}', \mathscr{P}')$.

Definition 2 Let $(\Omega, \mathscr{A}, \{P_\theta, \theta \in \Theta\})$ and $(\Omega', \mathscr{A}', \{P'_\theta, \theta \in \Theta\})$ be two statistical spaces having the same parameter and the same parameter space. We call

$$(\Omega \times \Omega', \mathscr{A} \otimes \mathscr{A}', \{P_\theta \otimes P'_\theta, \theta \in \Theta\})$$

the *weak product* of these two statistical spaces, which is also written as

$$(\Omega, \mathscr{A}, \{P_\theta, \theta \in \Theta\}) \times (\Omega', \mathscr{A}', \{P'_\theta, \theta \in \Theta\}).$$

In particular, we call the sample space associated with a sample of size n the weak product of n identical statistical spaces, i.e.,

$$(\Omega, \mathscr{A}, \mathscr{P})^n = (\Omega^n, \mathscr{A}^n, \{P^n, P \in \mathscr{P}\}).$$

Frequently, we shall refer to the latter space as the *sample.*

The notion of product corresponds in practice to stochastic independence of observations. Using Definitions 1 and 2, two cases can be considered. In the first case the parameters are not the same; in the second case they are assumed to be the same. The concept of sample of size n is very important and describes the statistical experiment in which one observes n times, independently and under the same conditions, a random element whose distribution is not completely known.

When the statistical spaces are dominated we can easily write the corresponding likelihood functions. With obvious notations we have

Definition 1 $\mathscr{L}(\omega, \omega'; \theta, \theta') = \mathscr{L}(\omega, \theta)\mathscr{L}'(\omega', \theta').$

Definition 2 $\mathscr{L}(\omega, \omega'; \theta) = \mathscr{L}(\omega, \theta)\mathscr{L}'(\omega', \theta).$

For a sample, it is useful to introduce

$$l(\omega, \theta) = \log \mathscr{L}(\omega, \theta)$$

and then we have

$$l(\omega_1, \ldots, \omega_n; \theta) = l(\omega_1, \theta) + \cdots + l(\omega_n, \theta).$$

Remark Using the concepts of product and weak product, we can define a mixed product when only a subset of the parameter is the same for the two statistical spaces involved. We do not introduce a special notation for such a product. Obviously, for every product, the statistics defined by the projections $(\omega, \omega') \to \omega$ and $(\omega, \omega') \to \omega'$ are independent.

Definition 3 Let $(\Omega, \mathscr{A}, \mathscr{P})^n$ be a sample of size n. For every point $(\omega_1, \ldots, \omega_n) \in \Omega^n$, the distribution on $(\Omega, \mathscr{A})$, defined by

$$\forall A \in \mathscr{A}, \qquad P_n(\omega_1, \ldots, \omega_n; A) = \frac{1}{n} \sum_{i=1}^{n} 1_A(\omega_i),$$

is called the *sample distribution.*

The strong law of large numbers and some other theorems of probability theory [3] show that if P is the common distribution of the random elements $\omega_1, \ldots, \omega_n$, the distribution P_n converges to P in a certain sense [3] when n increases to infinity. This fact is very useful in statistics. For example, to study concepts defined on $(\Omega, \mathscr{A}, \mathscr{P})$ such as moment, conditional distribution, etc., we use as statistic (defined on $(\Omega, \mathscr{A}, \mathscr{P})^n$) the corresponding concept related to $P_n(\omega_1, \ldots, \omega_n)$. For the statistic we generally use the same word as for the concept which is considered, but add the adjective *sample* (see Exercises 13 and 14). Note that sample moments always exist, but they are consistent estimators (Definition 6.1.2) if and only if the corresponding moment exists for every distribution belonging to $\mathscr{P}$.

Exercises

1. An urn contains N balls. Assume that N_i of these balls bear number i ($i = 1, \ldots, r$). The integers N_i are unknown but r and N are known and $N_1 + \cdots + N_r = N$. One draws n balls without replacement ($n < N$). What is the statistical space induced by the frequencies, i.e., the r-tuple, $(n_1 n_2, \ldots, n_r)$ where n_i is the number of balls drawn bearing the label i?

2. Answer the questions in Exercise 1 if the balls are drawn with replacement. Specialize to the case $r = 2$.

3. We have two independent sets of data and each of them can be considered as independent observations of one normal random variable. What is the corresponding statistical space?

4. Show that the statistical spaces obtained in the exercises above are dominated and write the corresponding likelihood functions.

5. Consider the statistical space

$$[\mathbb{R}, \mathscr{B}_{\mathbb{R}}; N(m, \sigma^2), m \in \mathbb{R}, \sigma \in \mathbb{R}^+]^2.$$

Show that the two statistics

$$(x_1, x_2) \to x_1 - x_2; \qquad (x_1, x_2) \to x_1 + x_2$$

are independent. Generalize this result.

6. We have one observation on a two-dimensional random vector having probability density function $f_\theta(x, y) = e^{-\theta x - (y/\theta)}$, $x, y \geqslant 0$, where θ is a positive and unknown parameter. Write the corresponding statistical space and find a statistic having zero expectation.

7. Consider a statistical space $[\Omega, \mathscr{A}, \mathscr{P}]$, where Ω is finite. Give a necessary and sufficient condition for the completeness of this statistical space. Apply the result to the statistical spaces obtained in Exercises 1 and 2.

8. Consider the following distribution P_θ on $(\mathbb{R}, \mathscr{B}_{\mathbb{R}})$:

$$P_\theta(\{-1\}) = \theta, \qquad P_\theta(\{n\}) = (1 - \theta)^2\theta^n, \qquad n = 0, 1, \ldots, \infty.$$

If $\theta \in]0, 1[$ is an unknown parameter and if we have one observation from P_θ, show that the corresponding statistical space is quasi-complete but not complete.

9. In n independent trials one event having probability p occurs v times. We assume that n and v are known but p is unknown. Show that the corresponding statistical space is complete.

10. Show that the following statistical spaces are complete:

(a) $(\{0, 1, \ldots, \infty\}, \mathscr{A}; \mathscr{P}(\lambda), \lambda \in \mathbb{R}^+)$, where $\mathscr{A}$ is the subfield of all subsets of $\{0, 1, \ldots, \infty\}$ and $\mathscr{P}(\lambda)$ is the Poisson distribution having parameter λ;

(b) $(\mathbb{R}, \mathscr{B}_{\mathbb{R}}; N(m, 1), m \in \mathbb{R})$;

(c) $(\mathbb{R}^+, \mathscr{B}_{\mathbb{R}^+}; [\theta^a/\Gamma(a)]e^{-\theta x}x^{a-1}, \theta \in \mathbb{R}^+)$,

where a is a given positive number.

11. A random variable has a Poisson distribution $\mathscr{P}(\theta)$ where the unknown parameter θ has a prior distribution whose probability density function is

$$[\lambda^a/\Gamma(a)]e^{-\theta\lambda}\theta^{a-1}, \qquad \theta > 0,$$

where a and λ are given positive numbers. What is the corresponding posterior distribution?

12. Consider the following statistical space

$$(\mathbb{R}, \mathscr{B}_{\mathbb{R}}; N(m, \sigma^2), m \in \mathbb{R}, \sigma \in \mathbb{R}^+)^n.$$

Find the statistical space induced by the statistic

$$(x_1, \ldots, x_n) \to \left(\bar{x} = \frac{x_1 + \cdots + x_n}{n}, S = \frac{1}{n}\sum_1^n (x_i - \bar{x})^2\right).$$

Is this latter statistical space a product of statistical spaces? Write the likelihood function corresponding to both statistical spaces.

13. Consider a sample of size n on a random variable having a continuous cumulative distribution function. Write the formulas which define the statistics, sample mean, sample variance, and sample median.

14. Consider a sample of size n from a k-dimensional random vector. Write formulas which define the statistics, sample correlation coefficient (between two components), sample covariance matrix, sample marginal cumulative distribution function (of one component), and the sample expectation of one component given another one.

CHAPTER **2**

Sufficiency and Freedom

The concepts of the previous chapter are straightforward adaptations to statistics of probability theory. We now introduce the notions of sufficiency and freedom,† which are basic and useful in statistics but do not correspond to any notion in probability theory.

1. Sufficient σ-Fields and Sufficient Statistics

Definition 1 Let $(\Omega, \mathscr{A}, \mathscr{P})$ be a statistical space. A subfield $\mathscr{B}$ of $\mathscr{A}$ is *sufficient* if and only if for every set $A \in \mathscr{A}$, a determination of $P(A \mid \mathscr{B})$ exists which is the same for all distributions $P \in \mathscr{P}$. This condition is equivalent to the following. For every integrable statistic X, a determination of $E_P(X \mid \mathscr{B})$ exists which is the same for all distributions $P \in \mathscr{P}$.

This equivalence is true for indicator functions. Then by linearity and monotone convergence (Property A.4.5) one can show that the equivalence is also true for every integrable statistic.

Note that the sufficiency of a subfield depends directly on the family $\mathscr{P}$; if $\mathscr{P}$ is extended, $\mathscr{B}$ does not necessarily remain sufficient.

Definition 2 We say that the statistic T, defined on the statistical space

† Although the concept of freedom can be found in R. A. Fisher's writing, elevating it to a basic concept parallel to sufficiency is due to Barra. (Ed.)

$(\Omega, \mathscr{A}, \mathscr{P})$ taking values in $(\mathscr{X}, \mathscr{C})$ is *sufficient* if and only if the subfield $T^{-1}(\mathscr{C})$ is sufficient.

In most cases one looks for sufficient statistics, but for theoretical purposes the concept of sufficient subfield is more convenient than the concept of sufficient statistic. A sufficient subfield can exist without having a sufficient statistic taking values in a given measurable space.

Theorem 1 *Let $(\mathbb{R}^n, \mathscr{B}_{\mathbb{R}^n}, \mathscr{P})$ be a statistical space. The vector (valued) statistic T is sufficient if and only if there exists a determination of the conditional distribution given T which is the same for every distribution belonging to $\mathscr{P}$.*

The sufficient condition is obvious. The necessary condition is established in [44, p. 48]. It is based on the fact that in this case a conditional distribution given T always exists.

This proposition emphasizes the fundamental concept of sufficiency. If we can write the statistical space as

$$(\mathscr{X} \times \mathscr{Y}, \mathscr{A} \otimes \mathscr{B}, \mathscr{P}),$$

the statistic $(x, y) \to y$ is sufficient if and only if the conditional distribution of x given y does not depend on the parameter. That is, when y is known, the value of x is of no importance in the problem. Therefore y summarizes all the information that the initial observations can give about the unknown parameter. When possible, one begins the study of a statistical problem by looking for a sufficient statistic. (See Exercise 1.4.)

The following two propositions are obvious.

Theorem 2 *If the subfield $\mathscr{B}$ is sufficient for the statistical space $(\Omega, \mathscr{A}, \mathscr{P})$, it is sufficient for the statistical space $(\Omega, \mathscr{A}, \mathscr{P}')$ if $\mathscr{P}'$ is included in $\mathscr{P}$ or if $\mathscr{P}'$ is the family of convex combinations of distributions belonging to $\mathscr{P}$.*

Theorem 3 *Every statistic which is equivalent to a sufficient statistic is sufficient.*

Theorem 4 *Let $\mathscr{B}$ be a sufficient subfield for the statistical space $(\Omega, \mathscr{A}, \mathscr{P})$ and $\mathscr{C}$ be a subfield of $\mathscr{B}$. Then $\mathscr{C}$ is sufficient for the statistical space $(\Omega, \mathscr{B}, \mathscr{P}_{\mathscr{B}})$ if and only if it is sufficient for the statistical space $(\Omega, \mathscr{A}, \mathscr{P})$.*

Proof We have written $\mathscr{P}_{\mathscr{B}}$ for the family of distributions belonging to $\mathscr{P}$ considered as distributions on $\mathscr{B}$. The sufficient condition is obvious. Property A.4.6 shows that

$$\forall P \in \mathscr{P}, \qquad \forall A \in \mathscr{A}; \qquad P(A \mid \mathscr{C}) = E_P(P(A \mid \mathscr{B}) \mid \mathscr{C}).$$

But $\mathscr{B}$ is sufficient. Thus $P(A|\mathscr{B})$ is a $\mathscr{B}$-measurable statistic and $E_P(P(A|\mathscr{B})|\mathscr{C})$ does not depend on P because $\mathscr{C}$ is sufficient for $(\Omega, \mathscr{B}, \mathscr{P}_{\mathscr{B}})$. ∎

2. Factorization Criterion of Sufficiency

When the given statistical space is dominated, the following theorem is fundamental.

Theorem 1 *Let* $(\Omega, \mathscr{A}, \{P_\theta, \theta \in \Theta\})$ *be a dominated statistical space and* P^* *a special probability distribution which dominates this space. The subfield* $\mathscr{B}$ *of* $\mathscr{A}$ *is sufficient if and only if there exists a determination of the densities* $p_\theta = dP_\theta/dP^*$ *which are* $\mathscr{B}$*-measurable for every* $\theta \in \Theta$*. Then for every set* $A \in \mathscr{A}$*, one can choose a determination of* $P^*(A|\mathscr{B})$ *as a common determination of the conditional probabilities* $P(A|\mathscr{B})$.

Corollary 1 *If* $\mathscr{B}$ *is a sufficient subfield for the dominated statistical space* $(\Omega, \mathscr{A}, \mathscr{P})$*, then every subfield* $\mathscr{B}'$*, where* $\mathscr{B} \subset \mathscr{B}' \subset \mathscr{A}$*, is sufficient.*

This corollary is a straightforward consequence of Theorem 1. Note that, if the statistical space is not dominated, this corollary is not always true.

Corollary 2 *Let* $\mathscr{B}$ *and* $\mathscr{B}'$ *be two subfields which are sufficient for the dominated statistical spaces* $(\Omega, \mathscr{A}, \mathscr{P})$ *and* $(\Omega', \mathscr{A}', \mathscr{P}')$*, respectively. Then* $\mathscr{B} \otimes \mathscr{B}'$ *is sufficient for the* $(\Omega, \mathscr{A}, \mathscr{P}) \otimes (\Omega', \mathscr{A}', \mathscr{P}')$*, the product of these statistical spaces.*

Proof The density $p_\theta(\omega)p'_{\theta'}(\omega')$ defined on the product $\Omega \times \Omega'$ is $\mathscr{B} \otimes \mathscr{B}'$-measurable if and only if $p_\theta(\omega)$ and $p'_{\theta'}(\omega')$ are $\mathscr{B}$ and $\mathscr{B}'$-measurable, respectively (see [53, p. 72]). ∎

Specifically, if the subfields $\mathscr{B}$ and $\mathscr{B}'$ are induced by the statistics T and T', respectively, then the statistic (T, T') is sufficient for the product of the given statistical spaces.

It is obvious that the same result is true for the weak product, but in this case, there often exists a better sufficient statistic (possibly of lower dimension).

Proof of Theorem 1 (a) *Necessity.* Suppose that the subfield $\mathscr{B}$ is sufficient. For every event $A \in \mathscr{A}$, let $P(A|\mathscr{B})$ be the determination of $P_\theta(A|\mathscr{B})$ which is the same for all θ. We have

$$\forall B \in \mathscr{B}, \quad \forall \theta \in \Theta, \qquad P_\theta(A \cap B) = \int_B P(A|\mathscr{B})\, dP_\theta. \tag{1}$$

Using a convex combination and Corollary A.4.1, we obtain

$$\forall B \in \mathscr{B}, \qquad P^*(A \cap B) = \int_B P(A \mid \mathscr{B})\, dP^*.$$

This shows that $P(A \mid \mathscr{B})$ is a determination of $P^*(A \mid \mathscr{B})$ and we can write (1) in the form,

$$\forall \theta \in \Theta, \qquad P_\theta(A) = \int P^*(A \mid \mathscr{B})\, dP_\theta^{\mathscr{B}},$$

where the restriction $P_\theta^{\mathscr{B}}$ of P_θ on $\mathscr{B}$ has, with respect to the restriction of P^* on $\mathscr{B}$, a density φ_θ which is obviously $\mathscr{B}$-measurable. Thus we have

$$\int P^*(A \mid \mathscr{B})\, dP_\theta^{\mathscr{B}} = \int P^*(A \mid \mathscr{B}) \varphi_\theta \, dP^*.$$

Using identity of conditional expectation (or Property A.4.8) we have

$$\int_A \varphi_\theta \, dP^* = \int P^*(A \mid \mathscr{B}) \varphi_\theta \, dP^*.$$

Finally

$$\forall \theta \in \Theta, \qquad \forall A \in \mathscr{A}, \qquad \int_A \varphi_\theta \, dP^* = P_\theta(A),$$

and we may choose φ_θ as a determination of the density dP_θ/dP^*.

(b) *Sufficiency.* Suppose that the determination p_θ of the density dP_θ/dP^* is $\mathscr{B}$-measurable. We must show that for every θ we can choose $P^*(A \mid \mathscr{B})$ as a determination of $P_\theta(A \mid \mathscr{B})$. We have

$$\forall A \in \mathscr{A}, \quad \forall B \in \mathscr{B}, \quad \forall \theta \in \Theta, \quad P_\theta(A \cap B) = \int 1_A \cdot 1_B \, dP_\theta = \int 1_A \cdot 1_B p_\theta \, dP^*.$$

But $1_B \cdot p_\theta$ is $\mathscr{B}$-measurable and using Property A.4.8 we have

$$P_\theta(A \cap B) = \int P^*(A \mid \mathscr{B}) 1_B p_\theta \, dP^* = \int_B P^*(A \mid \mathscr{B})\, dP_\theta. \qquad \blacksquare$$

Theorem 2 (Neyman) *Let $(\Omega, \mathscr{A}; \{p_\theta, \theta \in \Theta\})$ be a dominated statistical space. The statistic T taking values in $(\mathscr{T}, \mathscr{C})$ is sufficient, if and only if there exist*

(a) *a nonnegative real valued $\mathscr{A}$-measurable function h defined on Ω,*
(b) *a $\mathscr{C}$-measurable function g_θ defined on $\mathscr{T}$, such that for every $\theta \in \Theta$,*

$$\forall \theta \in \Theta, \qquad \forall \omega \in \Omega, \qquad p_\theta(\omega) = g_\theta(T(\omega)) h(\omega), \qquad \mu\text{-a.e.}$$

where μ dominates the given statistical space.

Proof Let P^* be a special distribution which dominates the given statistical space. Using Definition 1.2 and Theorem 1, we see that the statistic T is sufficient if and only if there exists a $T^{-1}(\mathscr{C})$-measurable determination of dP_θ/dP^*, or (by Lemma A.1.1) if and only if there exists a function g_θ defined on $(\mathscr{T}, \mathscr{C})$, which is $\mathscr{C}$-measurable for every $\theta \in \Theta$, where

$$\forall \theta \in \Theta, \qquad \forall \omega \in \Omega, \qquad \frac{dP_\theta}{dP^*}(\omega) = g_\theta(T(\omega)), \qquad \mu\text{-a.e.}$$

Moreover P^* is absolutely continuous with respect to μ (Theorem 1.1.1). Then, letting

$$dP^*/d\mu = h,$$

we have

$$p_\theta(\omega) = g_\theta(T(\omega))h(\omega) \qquad \mu\text{-a.e.}$$ ∎

Remark If the measures P_T^* and P_θ^T are the images of P^* and P_θ, induced by T on $(\mathscr{T}, \mathscr{C})$, then

$$g_\theta = dP_\theta^T/dP_T^*.$$

Now, Theorem A.1.1 yields

$$\forall C \in \mathscr{C}, \qquad P_\theta^T(C) = P_\theta(T^{-1}(C)) = \int_{T^{-1}(C)} dP_\theta$$

$$= \int_{T^{-1}(C)} g_\theta(T(\omega))\, dP^*(\omega) = \int_C g_\theta(t)\, dP_T^*.$$

3. Projection of a Statistic

Let $(\Omega, \mathscr{A}, \mathscr{P})$ be a statistical space, $\mathscr{B}$ a subfield of $\mathscr{A}$ and X an integrable statistic. In general, $E_P(X|\mathscr{B})$ depends on P and the conditional expectation of a statistic is not a statistic. Note the difference with probability theory where the conditional expectation of a random variable *is* a random variable. This emphasizes the property of sufficiency. If $\mathscr{B}$ is sufficient, $E_P(X|\mathscr{B})$ does not depend on P and thus defines a statistic.

Definition 1 Let $(\Omega, \mathscr{A}, \mathscr{P})$ be a statistical space. We say that the integrable statistic X has a *projection on the subfield* $\mathscr{B}$ *of* $\mathscr{A}$, if a determination of $E_P(X|\mathscr{B})$ exists which is the same for all distributions $P \in \mathscr{P}$. This common determination, called the *projection of* X *on* $\mathscr{B}$, is an integrable statistic and has the same image as X.

When there is no ambiguity, we write $E(X|\mathscr{B})$ for this common determination. We extend this definition to the case of a vector-valued statistic by applying this definition to each component.

Theorem 1 *Let $(\Omega, \mathscr{A}, \mathscr{P})$ be a statistical space, $\mathscr{B}$ a subfield of $\mathscr{A}$, and X an integrable statistic valued on $\mathbb{R}^k$ which has a projection on $\mathscr{B}$. Then, if $g(X)$ is an integrable statistic, for every real-valued function g defined on $\mathbb{R}^k$ which is convex and continuous, we have*

$$\forall P \in \mathscr{P}, \qquad E_P(g(X)) \geqslant E_P(g(E(X|\mathscr{B}))). \tag{1}$$

Proof Integrate both parts of the Jensen inequality (Property A.4.9) applied to g,

$$E_P(g(X)|\mathscr{B}) \geqslant g(E(X|\mathscr{B})).$$

∎

Theorem 2 (Rao–Blackwell) *Let $(\Omega, \mathscr{A}, \mathscr{P})$ be a statistical space, $\mathscr{B}$ a sufficient subfield and X a vector-valued statistic which has a covariance matrix Λ_X. Then the projection S of X on $\mathscr{B}$ has a covariance matrix Λ_S and*

$$\forall P \in \mathscr{P}, \qquad \Lambda_S \leqslant \Lambda_X.$$

(Covariance matrices are ordered by the corresponding quadratic forms).

Proof If $\mathbb{R}^s$ is the space of values of X,

$$\forall z \in \mathbb{R}^s, \qquad {}^t z \Lambda_X z = \sigma_P^2(\langle z | X \rangle), \qquad {}^t z \Lambda_S z = \sigma_P^2(\langle z | S \rangle).$$

Thus using Corollary A.4.4 we obtain

$$\forall z \in \mathbb{R}^s \qquad {}^t z \Lambda_S z \leqslant {}^t z \Lambda_X z.$$

∎

Definition 2 Let $(\Omega, \mathscr{A}, \mathscr{P})$ be a statistical space and $\mathscr{B}$ a subfield of $\mathscr{A}$. We say that the integrable statistic X is *conditionally mean-free* given $\mathscr{B}$ if $E_P(X|\mathscr{B})$ is a constant function on $\mathscr{P} \times \Omega$.

This notion of conditional mean-freedom is stronger than the notion of mean-freedom (Definition 1.2.7), except in the case where $\mathscr{B}$ is complete. The following theorem shows that in this case these two notions are equivalent.

Theorem 3 *Let $(\Omega, \mathscr{A}, \mathscr{P})$ be a statistical space and $\mathscr{B}$ a subfield of $\mathscr{A}$. The subfield $\mathscr{B}$ is complete if and only if every mean-free statistic which has a projection on $\mathscr{B}$ is conditionally mean-free given $\mathscr{B}$.*

Proof (a) *Necessity.* Let X be a mean-free statistic which has a projection on $\mathscr{B}$ and let

$$Y = E(X|\mathscr{B}) - E(X).$$

The statistic Y is integrable, $\mathscr{B}$-measurable and has zero expectation. Moreover, since $\mathscr{B}$ is complete, the statistic Y is $\mathscr{P}$-equivalent to zero. Then

$$E(X|\mathscr{B}) = E(X), \qquad \mathscr{P}\text{-a.e.}$$

That is, X is conditionally mean-free given $\mathscr{B}$.

(b) *Sufficiency.* Suppose that $\mathscr{B}$ is not complete. There exists a real valued statistic Y, not $\mathscr{P}$-equivalent to zero, which is $\mathscr{B}$-measurable, integrable, and has a zero expectation. Then, since Y is $\mathscr{B}$-measurable, it has a projection on $\mathscr{B}$ and

$$\forall P \in \mathscr{P}, \qquad E_P(Y|\mathscr{B}) = Y.$$

Thus we cannot have

$$E_P(Y|\mathscr{B}) = E(Y) = 0 \qquad \mathscr{P}\text{-a.e.}$$

∎

One can prove easily by the same argument

Theorem 4 *Let $(\Omega, \mathscr{A}, \mathscr{P})$ be a statistical space and $\mathscr{B}$ a subfield of $\mathscr{A}$. The subfield $\mathscr{B}$ is quasi-complete if and only if every bounded mean-free statistic which has a projection on $\mathscr{B}$ is conditionally mean-free given $\mathscr{B}$.*

Definition 3 Let $(\Omega, \mathscr{A}, \mathscr{P})$ be a statistical space and $\mathscr{L}$ a linear subspace of $L_1(\Omega, \mathscr{A}, \mathscr{P})$ (Definition 12.3.2). We say that the subfield $\mathscr{B}$ of $\mathscr{A}$ is *weakly sufficient* with respect to $\mathscr{L}$ if every statistic belonging to $\mathscr{L}$ has a projection on $\mathscr{B}$.

Some examples of application of this definition are given in Linnik [42]. Note the duality between sufficient and mean-free.

4. Free Subfields and Distribution-Free Statistics

We have introduced some concepts of freedom, i.e., of nondependence with respect to the unknown parameter in connection with the notion of sufficiency. The following definition complements these previous definitions.

Definition 1 If $(\Omega, \mathscr{A}, \mathscr{P})$ is a statistical space, we say that

(a) the event A is *free* if $P(A)$ is constant when P belongs to $\mathscr{P}$.
(b) the subfield $\mathscr{B}$ of $\mathscr{A}$ is *free* if every event belonging to $\mathscr{B}$ is free
(c) the statistic T, taking values in $(\mathscr{T}, \mathscr{C})$ is *distribution-free* if $T^{-1}(\mathscr{C})$ is free, i.e., if the distribution of T is the same for all distributions $P \in \mathscr{P}$.

For examples see Exercises 8–11.

The trivial subfield $(\varnothing, \Omega)$ is always free. If the statistical space $(\Omega, \mathscr{A}, \mathscr{P})$ is quasi-complete, only the $\mathscr{P}$-negligible events and their complements are free. If a statistical space is not quasi-complete, it is possible that all free events are trivial.

For freedom there is no criterion analogous to the factorization criterion for sufficiency. We shall see (Sections 7 and 12.4) that finding free events is difficult. For some classical examples, see Exercises 9–11.

5. $\mathscr{P}$-Minimum Sufficient Subfields and Statistics

We now look for sufficient subfields or statistics which correspond to the maximum reduction of the data. We note two steps in this process: first, finding a $\mathscr{P}$-minimum sufficient subfield, then (Section 12.2) finding among the statistics which induce the previous subfield, the one taking values in a linear space whose dimension is minimum. Note that the transitive property (Theorem 1.4) shows that if we obtain a "minimum" statistical space, every other "minimum" statistical space will be "equivalent."

Definition 1 The *minimum sufficient subfield* is the sufficient subfield, when it exists, which is included in every sufficient subfield.

If such a subfield exists, it is unique, since it is the intersection of all sufficient subfields. Note (remark after Theorem 3 below) that this definition is of no interest if Ω is not countable. This general case will be treated in Definition 4.

Definition 2 A sufficient statistic is said to be *minimum* if it induces the minimum sufficient subfield.

If X is a scalar minimum sufficient statistic and Y a sufficient statistic, there exists a measurable function such that

$$X = g(Y).$$

Then if Y is minimum, X and Y are equivalent.

Theorem 1 *Let $(\Omega, \mathscr{A}, \mathscr{P})$ be a statistical space and $\mathscr{N}$ the family of $\mathscr{P}$-negligible events. For every subfield $\mathscr{B}$ of $\mathscr{A}$, the subfield generated by $\mathscr{B} \cup \mathscr{N}$ is the subfield $\overline{\mathscr{B}}$ of events $\overline{B} \in \mathscr{A}$ for which there exists an event $B \in \mathscr{B}$ such that $\overline{B} \triangle B$ belongs to $\mathscr{N}$.*

Proof It is obvious that the family $\overline{\mathscr{B}}$ of sets such that

$$B \cup N - B \cap N = B \triangle N \qquad (B \in \mathscr{B},\ N \in \mathscr{N})$$

contains $\varnothing$ and Ω and is closed under complementation. On the other hand

$$\{\bigcup_i (B_i \triangle N_i)\} \triangle \{\bigcup_i B_i\} \subset \bigcup_i N_i.$$

Here $\bigcup_i B_i \in \mathscr{B}$ and $\bigcup_i N_i \in \mathscr{N}$ since this family is closed under countable unions. Then $\overline{\mathscr{B}}$ is a subfield and thus is the subfield generated by $\mathscr{B} \cup \mathscr{N}$.

Definition 3 We say that the subfields $\mathscr{B}$ and $\mathscr{B}'$ are *$\mathscr{P}$-equal* if

$$\mathscr{B} \subset \overline{\mathscr{B}'}, \qquad \mathscr{B}' \subset \overline{\mathscr{B}},$$

or, equivalently, if

$$\overline{\mathscr{B}'} = \overline{\mathscr{B}}.$$

The equivalence between these two conditions is easy to prove. It can also be stated that for every set B belonging to $\mathscr{B}$ (or $B' \in \mathscr{B}'$), there exists B' belonging to $\mathscr{B}$ (or $B \in \mathscr{B}$) such that $B \triangle B'$ belongs to $\mathscr{N}$.

Remark Some use $\mathscr{B} \subset \mathscr{B}'$, $\mathscr{P}$-a.e., in place of $\mathscr{B} \subset \overline{\mathscr{B}'}$. We do not use this notation which is perhaps convenient but can be ambiguous.

Theorem 2 *If $X \sim X'$, the subfields induced by X and X' are $\mathscr{P}$-equal.*

Proof Let $(\mathscr{X}, \mathscr{C})$ be the space of the values of X and X'. We have

$$\{B \in X^{-1}(\mathscr{C})\} \Leftrightarrow \{\exists C \in \mathscr{C} \text{ such that } B = X^{-1}(C)\}.$$

Since

$$\omega \in X^{-1}(C) \triangle X'^{-1}(C) \quad \Rightarrow \quad X(\omega) \neq X'(\omega),$$

we have

$$X^{-1}(C) \triangle X'^{-1}(C) \in \mathscr{N}. \qquad \blacksquare$$

Definition 4 The sufficient subfield $\mathscr{B}_0$ is *$\mathscr{P}$-minimum*, if for every sufficient subfield $\mathscr{B}$, $\mathscr{B}_0$ is included in $\overline{\mathscr{B}}$. This condition is equivalent to saying that $\overline{\mathscr{B}}_0$ is the smallest sufficient subfield including $\mathscr{N}$.

Proof If $\mathscr{B}$ is a sufficient subfield, $\overline{\mathscr{B}}$ is also sufficient since, if we have

$$C \in \overline{\mathscr{B}}, \qquad B \in \mathscr{B}, \qquad C \triangle B \in \mathscr{N},$$

we also have

$$P(A \cap C) = P(A \cap B), \qquad \forall A \in \mathscr{A}, \qquad \forall P \in \mathscr{P},$$

and

$$\left| \int_C P(A|\mathscr{B})\, dP - \int_B P(A|\mathscr{B})\, dP \right| \leqslant \int |1_C - 1_B|\, dP = 0.$$

Then $P(A|\mathscr{B})$ is a determination of $P(A|\mathscr{B})$ and $\mathscr{B}_0$ is included in every sufficient subfield including $\mathscr{N}$. In particular, $\mathscr{B}_0$ is included in the smallest of these subfields, namely $\overline{\mathscr{B}_0}$.

Definition 5 The *sufficient statistic* X is *$\mathscr{P}$-minimum*, if it induces a $\mathscr{P}$-minimum sufficient statistic.

Theorem 3 *Let $(\Omega, \mathscr{A}, \mathscr{P})$ be a dominated statistical space and P^* a special distribution which dominates it. A sufficient subfield $\mathscr{B}$ of $\mathscr{A}$ is $\mathscr{P}$-minimum if and only if $\mathscr{B}$ is $\mathscr{P}$-equal to the smallest subfield which makes measurable a determination of probability density functions of the family $\mathscr{P}$ with respect to P^*.*

Let $\mathscr{B}_0$ be the smallest subfield for which a determination of densities $p_\theta = dP_\theta/dP^*$ is measurable for every θ. Let $\mathscr{B}'_0$ be the subfield corresponding in the same way to another determination p'_θ of these densities. First we shall prove that $\mathscr{B}_0$ and $\mathscr{B}'_0$ are $\mathscr{P}$-equal. We have

$$\forall \theta \in \Theta, \qquad p_\theta = p'_\theta, \qquad P^*\text{-a.e.}$$

Then, using Theorem 2 and remembering that P^*-equivalence or equality is identical to $\mathscr{P}$-equivalence or equality, we have

$$\forall \theta \in \Theta, \qquad p_\theta^{-1}(\mathscr{B}_{\mathbb{R}}) \overset{P^*}{=} p_\theta'^{-1}(\mathscr{B}_{\mathbb{R}}).$$

Finally, we obtain

$$\forall \theta \in \Theta, \qquad \overline{p_\theta^{-1}(\mathscr{B}_{\mathbb{R}})} = \overline{p_\theta'^{-1}(\mathscr{B}_{\mathbb{R}})}$$

and thus

$$\overline{\mathscr{B}_0} = \overline{\mathscr{B}'_0}.$$

Proof of Theorem 3 Let $\mathscr{B}_0$ be a sufficient subfield. A determination p_θ of the densities exists which is $\mathscr{B}_0$-measurable. Let $\mathscr{B}'$ be the smallest subfield for which these densities are measurable. We have

$$\mathscr{B}' \subset \mathscr{B}_0$$

and then

$$\overline{\mathscr{B}'} \subset \overline{\mathscr{B}_0}.$$

If, moreover, $\mathscr{B}_0$ is $\mathscr{P}$-minimum sufficient we have

$$\mathscr{B}_0 \subset \overline{\mathscr{B}'}$$

and then

$$\overline{\mathscr{B}_0} \subset \overline{\mathscr{B}'}.$$

Finally

$$\overline{\mathscr{B}'} = \overline{\mathscr{B}_0}.$$

Let p'_θ be another determination of the densities and let $\mathscr{B}'_0$ be the smallest subfield for which they are measurable. Using the preliminary remark we obtain

$$\overline{\mathscr{B}'} = \overline{\mathscr{B}'_0}$$

and finally

$$\overline{\mathscr{B}_0} = \overline{\mathscr{B}'_0}.$$

Conversely, let $\mathscr{B}_0$ be the smallest subfield making a determination p_θ of the densities measurable and let $\mathscr{B}$ be a subfield $\mathscr{P}$-equal to $\mathscr{B}_0$. We shall prove that a determination p'_θ of $E_{P^*}(p_\theta | \mathscr{B})$ is a determination of $dP_\theta | dP^*$. We have

$$\forall \theta \in \Theta, \quad \forall A \in \mathscr{B}, \qquad P_\theta(A) = E_{P^*}(1_A p_\theta) = E_{P^*}(1_A p'_\theta) = \int_A p'_\theta \, dP^*.$$

Moreover, if C belongs to $\overline{\mathscr{B}}$, then A belonging to $\mathscr{B}$ exists such that

$$P^*(C \triangle A) = 0$$

and thus

$$P_\theta(C) = P_\theta(A) = \int_A p'_\theta \, dP^* = \int_C p'_\theta \, dP^*.$$

Finally p'_θ and p_θ are P^*-equivalent on $\overline{\mathscr{B}_0}$ and then on $\mathscr{A}$ since they are $\overline{\mathscr{B}_0}$-measurable. Therefore, $\mathscr{B}$ is sufficient. Let $\mathscr{B}'$ be another sufficient subfield. There exists a determination p''_θ of the densities which is $\mathscr{B}'$-measurable. Let $\mathscr{B}''$ be the smallest subfield making the functions p''_θ measurable. We have $\mathscr{B}'' \subset \mathscr{B}'$ and then $\overline{\mathscr{B}''} \subset \overline{\mathscr{B}'}$.

But previously we have proved that $\overline{\mathscr{B}_0} = \overline{\mathscr{B}''}$ and by hypothesis we have $\overline{\mathscr{B}} = \overline{\mathscr{B}_0}$, whence $\overline{\mathscr{B}} = \overline{\mathscr{B}''}$ and $\mathscr{B} \subset \overline{\mathscr{B}} \subset \overline{\mathscr{B}'}$. Then $\mathscr{B}$ is $\mathscr{P}$-minimum. ∎

Remark We have seen the important part played by the sufficient subfields obtained as the smallest subfields making a determination of densities measurable. In applied problems, a special determination (e.g., continuous) is frequently chosen. The previous theorem shows that the family of $\mathscr{P}$-minimum sufficient subfields is the family of subfields which are $\mathscr{P}$-equal to the smallest subfield $\mathscr{B}^*$ making the chosen determination of densities measurable. Thus there exists a minimum sufficient subfield if and only if the intersection of all subfields $\mathscr{P}$-equal to $\mathscr{B}^*$ is still sufficient. This will be true only if Ω is countable.

Now, without loss of generality, suppose that a determination of densities has been chosen. We look for a statistic inducing the smallest subfield making this determination of densities measurable.

Theorem 4 *Let V be a linear space generated by $\mathscr{F}$ a family of measurable mappings from $(\Omega, \mathscr{A})$ to $(\mathbb{R}, \mathscr{B}_{\mathbb{R}})$, and let $\{X_i, i \in I\}$ be a subset of V which generates V. The smallest subfield making all the functions belonging to $\mathscr{F}$ measurable is identical to the smallest subfield making all the functions $X_i\{i \in I\}$ measurable.*

Proof Let $\mathscr{A}_{\mathscr{F}}$ and $\mathscr{A}_I$ be the two subfields defined in the theorem. All functions of the family $\mathscr{F}$ are $\mathscr{A}_{\mathscr{F}}$-measurable and any element of V is a finite linear combination of functions of $\mathscr{F}$. Then, in particular, the functions X_i are $\mathscr{A}_{\mathscr{F}}$-measurable and $\mathscr{A}_{\mathscr{F}} \supset \mathscr{A}_I$. Conversely, any element of V, specifically one of the family $\mathscr{F}$, is a linear combination of functions X_i, and is $\mathscr{A}_I$-measurable. Therefore, $\mathscr{A}_{\mathscr{F}} \supset \mathscr{A}_I$. ∎

Theorem 5 *Let $(\Omega, \mathscr{A}, \{p_\theta, \theta \in \Theta\})$ be a dominated statistical space for which we have chosen a special distribution as dominating measure. Let ψ be a homeomorphism from $\mathbb{R}^+$ to $\mathbb{R}$ and let $\mathscr{L}$ be the linear space generated by the constant functions and the following family of functions on Ω*

$$g_\theta = \psi(p_\theta), \qquad \theta \in \Theta.$$

If $\{1; f_i, i \in I\}$ generates $\mathscr{L}$, the mapping defined by

$$\omega \in \Omega \to X(\omega) = \{f_i(\omega), i \in I\}$$

is a $\mathscr{P}$-minimum sufficient statistic.

Proof Using the definition of a homeomorphism of $\mathbb{R}^+$ onto $\mathbb{R}$ (see Section 12.2) we have

$$g_\theta^{-1}(\mathscr{B}_{\mathbb{R}}) = p_\theta^{-1}(\psi^{-1}(\mathscr{B}_{\mathbb{R}})) = p_\theta^{-1}(\mathscr{B}_{\mathbb{R}^+}), \qquad \forall \theta \in \Theta.$$

Then the smallest subfield making the densities p_θ-measurable is identical to the subfield making the functions g_θ-measurable. Therefore, using Theorem 4, this subfield is also identical to the subfield making X measurable since a constant is always measurable and we have chosen the product subfield on $\mathbb{R}^I$. ∎

Let $\mathscr{L}'$ be the linear space generated by the functions g_θ. If the function 1 belongs to $\mathscr{L}'$, it is obvious that $\mathscr{L} = \mathscr{L}'$ and we verify that the constant function is not included in the $\mathscr{P}$-minimum sufficient statistic. But if the function 1 does not belong to $\mathscr{L}'$, the family $\{f_i\}$ generates $\mathscr{L}'$. The same is true for the following theorem and shows that each of Theorems 5 and 6 includes both cases.

Theorem 6 *Let $(\Omega, \mathscr{A}, \{p_\theta, \theta \in \Theta\})^n$ be a dominated sample in which there exists $\theta_0 \in \Theta$ such that $p_{\theta_0}(\omega) > 0$, $\forall \omega \in \Omega$. If the family $\{1; f_i(\omega), i \in I\}$ generates the linear space $\mathscr{L}$ generated by constant functions and functions*

$$\left\{\log\left(\frac{p_\theta(\omega)}{p_{\theta_0}(\omega)}\right), \theta \in \Theta\right\},$$

then the statistic T defined by

$$\forall \omega_j \in \Omega, \quad j = 1, \ldots, n, \qquad T(\omega_1, \ldots, \omega_n) = \left\{\sum_{j=1}^{n} f_i(\omega_j), i \in I\right\}$$

is $\mathscr{P}$-minimum sufficient.

Proof Note that $(P_{\theta_0})^n$ is a special distribution which dominates the sample. Let $\mathscr{L}_n$ be the linear space generated by the constant functions and by the family of functions on Ω^n depending on $\theta \in \Theta$ and defined by

$$\forall \omega_j \in \Omega, \quad j = 1, \ldots, n, \qquad \log \frac{p_\theta(\omega_1, \ldots, \omega_n)}{p_{\theta_0}(\omega_1, \ldots, \omega_n)} = \sum_{j=1}^{n} \log\left(\frac{p_\theta(\omega_j)}{p_{\theta_0}(\omega_j)}\right).$$

By hypothesis we have

$$\log(p_\theta(\omega)/p_{\theta_0}(\omega)) = c_\theta + \sum_{i \in I} c_\theta^i f_i(\omega), \tag{1}$$

where only a finite number of coefficients c_θ^i are not zero. By definition, a function belonging to $\mathscr{L}_n$ can be written as

$$f(\omega_1, \ldots, \omega_n) = a + \sum_k a_k \log \frac{p_{\theta_k}(\omega_1, \ldots, \omega_n)}{p_{\theta_0}(\omega_1, \ldots, \omega_n)} = a + \sum_{k,j} a_k \log\left(\frac{p_{\theta_k}(\omega_j)}{p_{\theta_0}(\omega_j)}\right).$$

Using (1), we obtain

$$\begin{aligned} f(\omega_1, \ldots, \omega_n) &= a + \sum_k a_k c_{\theta_k} + \sum_{k,j,i} a_k c_{\theta_k}^i f_i(\omega_j), \\ &= a + \sum_k a_k c_{\theta_k} + \sum_{i \in I} \left(\sum_k a_k c_{\theta_k}^i\right)\left(\sum_{j=1}^{n} f_i(\omega_j)\right). \end{aligned}$$

It is easy to see that the functions

$$\left\{\sum_{j=1}^{n} f_i(\omega_j),\ i \in I\right\}$$

belong to $\mathscr{L}_n$ and together with the function 1, these functions generate $\mathscr{L}_n$. We can now apply Theorem 5 with the logarithmic function as the function ψ. ∎

Note that Theorems 4–6 are still true if V has a topology in which V is generated by a countable subfamily.

Remark When Ω is a finite-dimensional linear space, Theorem 6 gives an interesting result only if

$$n \dim \Omega \geqslant \dim \mathscr{L} - 1.$$

On the other hand, even if we use a basis of $\mathscr{L}$, it is not necessarily true that the sufficient statistic that we obtain will be of minimum dimension (See Section 12.2).

Theorem 7 *Let* $(\Omega, \mathscr{A}, \mathscr{P})$ *be a dominated statistical space. If a subfield* $\mathscr{B}$ *is sufficient and quasi-complete, then* $\mathscr{B}$ *is* $\mathscr{P}$*-minimum sufficient.*

Proof If $\mathscr{B}^*$ is the smallest subfield making a $\mathscr{B}$-measurable determination of the densities measurable, we have $\mathscr{B} \supset \mathscr{B}^*$. Moreover, for every A belonging to $\mathscr{A}$, the statistic $P(A \mid \mathscr{B}) - P(A \mid \mathscr{B}^*)$ is bounded, $\mathscr{B}$-measurable, and has zero expectation. Since $\mathscr{B}$ is quasi-complete, this statistic is $\mathscr{P}$-equivalent to zero. Specifically for every B belonging to $\mathscr{B}$ we have

$$1_B = P(B \mid \mathscr{B}^*), \qquad \mathscr{P}\text{-a.e.} \tag{2}$$

Let B' be the set on which $P(B \mid \mathscr{B}^*)$ is equal to 1. Using (2), $B' \triangle B$ is $\mathscr{P}$-negligible. Therefore, $\mathscr{B}$ is $\mathscr{P}$-equal to the subfield $\mathscr{B}^*$ and finally, $\mathscr{B}$ is $\mathscr{P}$-minimum. ∎

6. Relationship among Freedom, Completeness, Sufficiency, and Stochastic Independence

We have already noted some of these relations. The following results show that in many cases, two conditions chosen among freedom, sufficiency, and stochastic independence, imply the third. On the other hand, one can see in [58] that freedom and sufficiency can be considered as dual concepts.

Theorem 1 *Let $\mathscr{B}$ be a sufficient and quasi-complete subfield for the statistical space $(\Omega, \mathscr{A}, \mathscr{P})$ and let $\mathscr{C}$ be a free subfield of $\mathscr{A}$. Then $\mathscr{B}$ and $\mathscr{C}$ are independent for every distribution $P \in \mathscr{P}$.*

Proof Let C be an event of $\mathscr{C}$. By Theorem 3.3, we then have

$$P(C \mid \mathscr{B}) = P(C), \qquad \forall P \in \mathscr{P},$$

and

$$\forall B \in \mathscr{B}, \qquad P(B \cap C) = \int_B P(C \mid \mathscr{B})\, dP = P(C)P(B), \qquad \forall P \in \mathscr{P}.$$

Thus the subfields $\mathscr{B}$ and $\mathscr{C}$ are independent.

Corollary 1 *Let $(\Omega, \mathscr{A}, \mathscr{P})$ be a statistical space and T a sufficient statistic taking values in $(\mathscr{T}, \mathscr{C})$. If the statistical space $(\mathscr{T}, \mathscr{C}, \mathscr{P}_T)$ is quasi-complete, every free statistic defined on $(\Omega, \mathscr{A}, \mathscr{P})$ is independent of T.*

Proof The quasi-completeness of $(\mathscr{T}, \mathscr{C}, \mathrm{P}_T)$ implies the quasi-completeness of $(\Omega, T^{-1}(\mathscr{C}), \mathscr{P})$. ∎

For applications of this theorem see Exercise 14 and [42, pp. 10, 11].

Theorem 2 *Let $\mathscr{B}$ be a sufficient subfield for the statistical space $(\Omega, \mathscr{A}, \mathscr{P})$. Suppose that there do not exist two distributions $P \in \mathscr{P}$ and $P' \in \mathscr{P}$ and an event $C \in \mathscr{A}$ such that $P(C) = 1$, $P'(C) = 0$. If a subfield is independent of $\mathscr{B}$ for every distribution P belonging to $\mathscr{P}$, then this subfield is free.*

Proof Let $\mathscr{C}$ be a subfield independent of $\mathscr{B}$ for every distribution belonging to $\mathscr{P}$. For every event $C \in \mathscr{C}$ let B_P be the event of $\mathscr{B}$ defined by

$$P(C \mid \mathscr{B}) = P(C).$$

Using Property A.4.7 we have

$$P(B_P) = 1.$$

Now let P' be another distribution belonging to $\mathscr{P}$. Since $\mathscr{B}$ is sufficient, we have

$$P(C \mid \mathscr{B}) = P'(C \mid \mathscr{B}),$$

and then

$$B_P \cap B_{P'} \neq \varnothing \quad \Rightarrow \quad P(C) = P'(C).$$

But, by hypothesis,

$$\forall P, \quad P' \in \mathscr{P}, \qquad B_P \cap B_{P'} \neq \varnothing.$$ ∎

Theorem 3 *Let $\mathscr{C}$ be a free subfield for the statistical space $(\Omega, \mathscr{A}, \mathscr{P})$*

and $\mathscr{B}$ a subfield of $\mathscr{A}$, such that $\mathscr{C} \cup \mathscr{B}$ generates $\mathscr{A}$. Suppose that for every distribution $P \in \mathscr{P}$, $\mathscr{C}$ and $\mathscr{B}$ are independent. Then $\mathscr{B}$ is sufficient.

Proof The same argument that was used at the beginning of the previous proof shows that

$$\forall C \in \mathscr{C}, \qquad P(C \mid \mathscr{B}) = P(C), \qquad P\text{-a.e.}$$

Then, since $\mathscr{C}$ is free, C is conditionally free given $\mathscr{B}$. Therefore we have

$$\forall B \in \mathscr{B}, \quad \forall C \in \mathscr{C}, \qquad P(B \cap C \mid \mathscr{B}) = 1_B P(C \mid \mathscr{B}) = 1_B P(C),$$

i.e., $B \cap C$ is free given $\mathscr{B}$.

Let us consider the Boolean algebra $\mathscr{A}_0$ of events A such that

$$A = \bigcup_{i=1}^{n} [C_i \cap B_i], \qquad C_i \in \mathscr{B}, \quad B_i \in \mathscr{B}, \quad i = 1, \ldots, n,$$

where the sets $C_i \cap B_i$ are disjoint. We have shown that A is conditionally free given $\mathscr{B}$ and this property remains true by monotone convergence. Let $P \in \mathscr{P}$ and $P_0 \in \mathscr{P}$ be two distributions and let $B \in \mathscr{B}$ be an event such that $P(B) > 0$. Then, the two distributions on $\mathscr{A}$ defined by

$$A \in \mathscr{A}, \qquad P_B(A) = P(A \cap B)/P(B), \qquad P'_B(A) = \int_B P_0(A \mid \mathscr{B})\, dP/P(B),$$

have the same values on $\mathscr{A}_0$ and thus on $\mathscr{A}$ which is generated by $\mathscr{A}_0$. Thus we have

$$\forall B \in \mathscr{B}, \qquad \forall A \in \mathscr{A}, \qquad P(A \cap B) = \int_B P_0(A \mid \mathscr{B})\, dP. \qquad \blacksquare$$

In the following chapter, we discuss applications of the previous two theorems to the concept of information.

7. Existence of Free Events

This problem is difficult but important, as free events can be used for nonparametric tests (see Section 5.6) and for set estimation (see Section 6.3). There are few existence theorems. The best known is the following theorem of Liapounov. (For a proof see Theorem 12.4.2.)

Theorem 1 *Let $P_1, \ldots, P_N$ be N nonatomic distributions on $(\Omega, \mathscr{A})$. Then for every α belonging to $[0, 1]$, there exists an event A of $\mathscr{A}$ such that*

$$P_i(A) = \alpha, \qquad \forall i = 1, \ldots, N.$$

Theorem 2 *Let $\mathscr{B}$ be a sufficient subfield on the statistical space $(\Omega, \mathscr{A}, \mathscr{P})$ dominated by the special distribution P^*. If an event A exists such that*

$$P^*(A \mid \mathscr{B}) = \alpha, \qquad (\alpha \in]0, 1[),$$

then

$$P(A) = \alpha, \qquad \forall P \in \mathscr{P}. \tag{1}$$

Proof The subfield $\mathscr{B}$ is sufficient and we have

$$P(A \mid \mathscr{B}) = P^*(A \mid \mathscr{B}) = \alpha, \qquad \forall P \in \mathscr{P}.$$

Integration leads to (1). ∎

We note that Theorem 2 is useful only if $\mathscr{B}$ is strictly included in $\mathscr{A}$. Otherwise the event A does not exist. However, if $\mathscr{B}$ is quasi-complete, Theorem 3.3 shows that a free event having probability α exists if and only if an event A exists such that

$$P^*(A \mid \mathscr{B}) = \alpha.$$

In addition, this theorem reduces the problem of free events to the study of P^*. If P^* does not have $\mathscr{B}$-atoms there exists an event A such that

$$P^*(A \mid \mathscr{B}) = \alpha.$$

(For a proof see Soler [58; p. 86]).
We easily deduce the following

Corollary 1 *Consider the following dominated statistical space*

$$\left(\Omega, \mathscr{A}, \left\{P_\theta = \sum_{i=1}^{N} c_i(\theta) P_\theta^i, \theta \in \Theta\right\}\right).$$

For every $1 = 1 \cdots N$, let P_i^* be a special distribution which dominates the statistical space

$$(\Omega, \mathscr{A}, \{P_\theta^i, \theta \in \Theta\})$$

and let $\mathscr{B}_i$ be a sufficient subfield for this statistical space. If an event $A \in \mathscr{A}$ exists such that

$$P_i^*(A \mid \mathscr{B}_i) = \alpha, \qquad \forall i = 1, \ldots, N,$$

then

$$P_\theta(A) = \alpha, \qquad \forall \theta \in \Theta.$$

More generally, Linnik [41] and Kagan [35], with regard to the Behrens–

Fisher problem, have also studied the family of distributions of the form

$$P_\theta = \sum_1^N c_i(\theta)\mu_\theta^i,$$

where the positive measures μ_θ^i on $(\Omega, \mathscr{A})$ are such that there exists a countable and measurable subdivision of Ω

$$\Omega = \bigcup_n \Omega_n$$

such that

$$0 < \mu_\theta^i(\Omega_n) < \infty, \qquad \forall i, \theta, n.$$

Let $P_{\theta,n}^i$ be the distribution defined by

$$A \in \mathscr{A}, \qquad P_{\theta,n}^i(A) = \frac{\mu_\theta^i(A \cap \Omega_n)}{\mu_\theta^i(\Omega_n)},$$

and let $\mathscr{B}_i^n$ be the subfield consisting of intersections of Ω_n and events belonging to $\mathscr{B}_i$. Suppose that for every n, the statistical space

$$(\Omega_n, \mathscr{A} \cap \Omega_n, \{P_{\theta,n}^i, \theta \in \Theta, i = 1, \ldots, N\})$$

and the subfield $\mathscr{B}_i^n$ satisfy the assumptions of the colorollary. Then, for every integer n, there exists a set A_n of Ω_n such that

$$P_{\theta,n}^i(A_n) = \alpha, \qquad \forall \theta \in \Theta, \quad \forall i = 1, \ldots, N.$$

We conclude easily that

$$\forall \theta \in \Theta, \qquad P_\theta(\bigcup_n A_n) = \alpha.$$

Exercises

1. Consider a sample of size n on a real random variable having a distribution symmetric about zero. Write the corresponding statistical space and show that the sequence of ordered absolute values of observations define a sufficient statistic.

2. Find sufficient statistics on the following statistical spaces (in (a) and (b) $\mathscr{A}$ represents the subfield of all subsets of Ω).

(a) $[\{0, 1, \ldots, N\}, \mathscr{A}; \mathscr{B}(N, p, 1 - p), p \in [0, 1]]^n$;
(b) $[\{0, 1, \ldots, \infty\}, \mathscr{A}; \mathscr{P}(\lambda), \lambda \in \mathbb{R}^+]^n$;
(c) $[\mathbb{R}, \mathscr{B}_{\mathbb{R}}; N(m, 1), m \in \mathbb{R}]^n$;
(d) $[\mathbb{R}, \mathscr{B}_{\mathbb{R}}; N(0, \sigma^2), \sigma \in \mathbb{R}^+]^n$;
(e) $[\mathbb{R}^+, \mathscr{B}_{\mathbb{R}^+}; \theta^a e^{-\theta x} x^{a-1}/\Gamma(a), a \in \mathbb{R}^+, \theta \in \mathbb{R}^+]^n$;

(f) $[\mathbb{R}, \mathscr{B}_{\mathbb{R}}; N(m, \sigma^2), m \in \mathbb{R}, \sigma \in \mathbb{R}^+]^n$;
(g) $[\mathbb{R}^k, \mathscr{B}_{\mathbb{R}^k}; N(m, \Lambda), m \in \mathbb{R}^k, \Lambda \in \mathscr{S}_k^+]^n$.

3. Consider a sample of size n from the uniform distribution on $[\theta, \theta + \mu]$, where θ and $\mu > 0$ are unknown parameters. Show that the greatest and smallest observations define a sufficient statistic.

4. Let E be a finite set and let Ω be a subset of the family S of all distributions on E.

(a) If we have one observation on a random variable on E having an unknown distribution belonging to Ω, what is the corresponding statistical space? Find the condition under which a nontrivial sufficient statistic exists.

(b) We have a sample of size n on the same random variable. In the case $\Omega = S$ find a minimal sufficient statistic and the statistical space induced by this statistic. Compute the expectation and the covariance matrix of this statistic. In the case $\Omega \subset S$, find conditions for the existence of a sufficient statistic whose dimension is less than that of the previous sufficient statistic.

5. Let X be a random variable on $[0, 1]$ having a probability density function f and let f_θ be the probability density function of the random variable $Y = \theta X$. We have a sample of size n on Y and the unknown parameter θ is assumed to be positive. Write the corresponding statistical space and find f such that for every n the greatest observation is a sufficient statistic.

6. For every θ belonging to $]0, 1[$, compute the constant K_θ such that the function of integer k

$$k \to K_\theta \theta^k / k!$$

defines a distribution L_θ on the positive integers. Compute the expectation, the variance and the generating function of L_θ. Consider a sample of size n from distribution L_θ, where θ is unknown. Write the corresponding statistical space and find a sufficient statistic. Compute the statistical space induced by this statistic and show that this statistical space is complete. Finally, find a statistic having the identity as image.

7. We have a sample of size n from the bivariate normal distribution having expectation and covariance

$$\begin{pmatrix} m_1 \\ m_2 \end{pmatrix} \quad \text{and} \quad \begin{pmatrix} \sigma_1^2 & \rho\sigma_1\sigma_2 \\ \rho\sigma_1\sigma_2 & \sigma_2^2 \end{pmatrix},$$

respectively. Consider all possible cases in which some of the five parameters are known (the others being unknown). For each case find a sufficient statistic and compare its dimension to the number of unknown parameters.

8. Find some statistical spaces which are not quasi-complete and for which there does not exist a nontrivial free set. Find some statistical spaces admitting several free subfields. *Hint:* Consider statistical spaces for which the number of elements in Ω is 3 or 4.

9. Consider the statistical space

$$[\mathbb{R}, \mathscr{B}_{\mathbb{R}}; N(m, \sigma^2), m \in \mathbb{R}, \sigma \in \mathbb{R}^+]^n.$$

Show that Geary's statistic

$$G = (1/S^\alpha) \sum_1^n |x_i - \bar{x}|^{2\alpha},$$

where α is a given number, $n\bar{x} = \sum_1^n x_i$, and $S = \sum_1^n (x_i - \bar{x})^2$ is a free statistic which is independent of the statistic $(\bar{x}, S)$.

10. Show that if we consider a sample from a bivariate normal distribution whose correlation coefficient is known (the other parameters being unknown), the sample correlation coefficient is a free statistic.

11. Consider a sample of size $m + n$ on a random variable having a continuous cumulative distribution function. Let $F_n(x)$ be the sample cumulative distribution function of the first n observations and let $F_m(x)$ be the sample cumulative distribution function of the last m observations. Show that the Smirnov statistic

$$K = \sup_x |F_n(x) - F_m(x)|$$

is free.

12. Consider a sample of size n from the Cauchy distribution on $(\mathbb{R}, \mathscr{B}_{\mathbb{R}})$ having the probability density function

$$\frac{dP_\theta}{dx} = \frac{\theta}{\pi(x^2 + \theta^2)}, \qquad \theta \in \mathbb{R}^+,$$

where θ is an unknown parameter. Show that the subfield defined by symmetric Borel sets of $\mathbb{R}^n$ is a $\mathscr{P}$-minimum sufficient subfield.

13. Let $(\Omega, \mathscr{A}, \mathscr{P})$ be a statistical space and X a $\mathscr{P}$-minimum sufficient statistic which induces the statistical space $(\mathscr{X}, \mathscr{C}, \mathscr{P}_X)$. Show that $\mathscr{C}$ is a $\mathscr{P}_X$-minimum sufficient subfield for the latter statistical space.

14. Consider a sample $(x_i, x_i'; i = 1, \ldots, n)$ of size n on a bivariate normal distribution and let $\bar{x}, \bar{x}', S, S'$ be the sample expectations and the sample variances of each component. Find conditions under which the statistic

$$T = (SS')^{-r/2} \sum_{i=1}^n |x_i - \bar{x}|^r |x_i' - \bar{x}'|^r$$

is free and independent of the statistic $(\bar{x}, \bar{x}', S, S')$.

CHAPTER 3

Statistical Information

The concept of amount of *information* with respect to an unknown parameter, like sufficiency and freedom, is defined for a statistical space independently of the decision problem which will be studied in Chapter 4. We shall consider the most famous definition of statistical information, the one due to Fisher. However, other definitions due to Shannon and Kullback have been suggested (see, e.g., Kullback [37]).

1. Introduction

First we look for properties that we feel information should satisfy and then show that we cannot satisfy all intuitively desirable properties.

Let $(\Omega, \mathscr{A}; P_\theta, \theta \in \Theta)$ be a statistical space and I be an ordered group. If we want to represent the amount of information† included in a statistical space, by an element of I, intuition suggest that we assume the following:

(1) The information given by a statistic is equal to the information included in the statistical space induced by this statistic.

(2) The information given by a statistic is less than or equal to the information included in the statistical space on which this statistic is defined.

† Although the correct terminology is *amount of information*, for typographical convenience we shall use the term *information*.

(3) The information given by a sufficient statistic is equal to the information included in the statistical space on which this statistic is defined.

(4) The information given by a free statistic is equal to zero.

(5) Two equivalent statistics give the same information.

(6) The information given by two independent statistics is the sum of the informations given by each of them.

Property (5) shows that it is convenient to define information in a subfield $\mathscr{B}$ of $\mathscr{A}$ and we write this information as $\mathscr{I}(\mathscr{B})$. Then the first property states that a definition of information given by a statistic is equal to the information in the subfield induced by this statistic. Thus, we may consider the following properties

I_1: If the subfield $\mathscr{B}$ is free, $\mathscr{I}(\mathscr{B}) = 0$.

I_1': If $\mathscr{I}(\mathscr{B}) = 0$, the subfield $\mathscr{B}$ is free.

I_2: If $\mathscr{B} \subset \mathscr{A}$, then $\mathscr{I}(\mathscr{A}) \geqslant \mathscr{I}(\mathscr{B})$.

I_3: If $\mathscr{B}$ is sufficient, then $\mathscr{I}(\mathscr{B}) = \mathscr{I}(\mathscr{A})$.

I_3': If $\mathscr{B} \subset \mathscr{A}$ and $\mathscr{I}(\mathscr{B}) = \mathscr{I}(\mathscr{A})$, then $\mathscr{B}$ is sufficient.

I_4: If $\mathscr{B}$ and $\mathscr{B}'$ are independent subfields for every value of the unknown parameter and generate the subfield $\mathscr{B}''$ we have

$$\mathscr{I}(\mathscr{B}) + \mathscr{I}(\mathscr{B}') = \mathscr{I}(\mathscr{B}'').$$

Note that we have introduced Properties I_1' and I_3', the converses of I_1 and I_3 respectively; these are as natural as the other properties. However, we have seen (Section 2.6) that all these properties cannot be satisfied simultaneously. For instance, Properties I_1, I_3', I_4, imply that if $\mathscr{B}$ is free and independent of $\mathscr{B}'$ and if $\mathscr{B}$ and $\mathscr{B}'$ generate $\mathscr{A}$, then $\mathscr{B}$ is sufficient. This is Theorem 2.6.3. However, Properties I_1', I_2, I_3, I_4, imply that if $\mathscr{B}$ is sufficient and independent of $\mathscr{B}'$, then $\mathscr{B}'$ is free. But Theorem 2.6.2 shows that this property is not always true. One can construct examples (see Basu [2], Linnik [42, p. 5]) for which the properties are inconsistent.

Note also that the previous properties suggest the notion of conditional information. More precisely, suppose that for every subfields $\mathscr{B}$ and $\mathscr{B}'$ such that

$$\mathscr{B}' \subset \mathscr{B} \subset \mathscr{A}$$

we associate $\mathscr{I}(\mathscr{B} \mid \mathscr{B}')$, as the conditional information in $\mathscr{B}$ given $\mathscr{B}'$. If we identify $\mathscr{I}(\mathscr{B})$ with $\mathscr{I}(\mathscr{B} \mid \{\varnothing, \Omega\})$ we are led to consider Property

I_5: If $\mathscr{B}' \subset \mathscr{B} \subset \mathscr{A}$, then $\mathscr{I}(\mathscr{B}) = \mathscr{I}(\mathscr{B}') + \mathscr{I}(\mathscr{B} \mid \mathscr{B}')$,

as a definition of $\mathscr{I}(\mathscr{B} \mid \mathscr{B}')$. Then Property I_2 is equivalent to

I_2': $\mathscr{I}(\mathscr{B} \mid \mathscr{B}') \geqslant 0$, and property I_3 becomes

I_6: If $\mathscr{B}$ is sufficient, then $\mathscr{I}(\mathscr{A}\,|\,\mathscr{B}) = 0$.

Furthermore, consider the following property, which is stronger than I_1':

I_7: $\mathscr{I}(\mathscr{B}|\mathscr{B}') = 0$ $(\mathscr{B} \supset \mathscr{B}')$, implies that $\mathscr{B}$ is conditionally free given $\mathscr{B}'$, i.e., every function indicator of a set belonging to $\mathscr{B}$ has a projection on $\mathscr{B}'$.

If Property I_7 is satisfied, then we obtain the definition of sufficiency. Thus, we see that the condition, $\mathscr{I}(\mathscr{B}|\mathscr{B}') = 0$ $(\mathscr{B} \supset \mathscr{B}')$, can be considered as a relation of duality on pairs of subfields; sufficiency and freedom are the two parts of this duality.

In the usual problems, involving dominated statistical spaces, the classical definitions of information are linear functionals of the function $\log p_\theta(\omega)$, where p_θ is the probability density. For such notions of information, we can verify Properties I_4 and I_3 easily, by means of the factorization theorem of Neyman.

2. Information (according to Fisher)

Definition 1 Let

$$(\Omega, \mathscr{A}; dP_\theta/dv = p_\theta, \theta \in \Theta)$$

be a dominated statistical space where Θ is a subset of $\mathbb{R}^s$. If the random vector V_θ taking values in $(\mathbb{R}^s, \mathscr{B}_{\mathbb{R}^s})$ and defined on $(\Omega, \mathscr{A}, P_\theta)$ by

$$\forall \omega \in \Omega, \qquad V_\theta(\omega) = \operatorname{grad}_\theta \log p_\theta(\omega),$$

exists for every θ, has zero expectation, and admits a covariance matrix, then the *information*† $\mathscr{I}$, as a function of θ, is defined for every θ by the covariance matrix of V_θ.

Under general regularity conditions the information exists, as can be seen by the following argument. If the subset Ω_0 of Ω, defined by

$$p_\theta(\omega) > 0 \quad \Leftrightarrow \quad \omega \in \Omega_0$$

does not depend on θ, and if we can differentiate the following integral

† Since the only definition of information used in this book is that due to Fisher, hereafter we shall just use the term *information* instead of *information according to Fisher*.

under the integral sign,

$$\int_{\Omega_0} p_\theta(\omega)\, dv = 1,$$

we obtain

$$\int_{\Omega_0} \operatorname{grad}_\theta p_\theta(\omega)\, dv = 0,$$

and then

$$E(V_\theta) = \int_{\Omega_0} \operatorname{grad}_\theta \log p_\theta(\omega)\, p_\theta(\omega)\, dv = 0.$$

That is, V_θ exists and has zero expectation.

We observe that, when the information exists, it does not depend on the measure v which dominates the statistical space, since

$$\frac{dP_\theta}{dv'} = \frac{dP_\theta}{dv}\frac{dv}{dv'}$$

and the vector V_θ is the same whether measure v or measure v' is used. Finally, we shall consider that the information matrices are ordered by their associated quadratic forms.

Definition 2 Let $(\Omega, \mathscr{A}; P_\theta, \theta \in \Theta)$ be a dominated statistical space and let X be a statistic taking values in $(\mathscr{X}, \mathscr{C})$. If the statistical space induced by X admits an information matrix $\mathscr{I}_X$, we define the *information associated with* X by this matrix.

Definition 2 implies the following three theorems.

Theorem 1 *The statistic* X *is free if and only if* $\mathscr{I}_X = 0$.

Proof Let $p_\theta^X(x)$ be the probability density function of X and let V_θ^X be the random vector,

$$V_\theta^X(x) = \operatorname{grad}_\theta \log p_\theta^X(x).$$

We can easily see that if X is free, V_θ^X is equal to zero. Conversely, if $\mathscr{I}_X$ is equal to zero, since V_θ^X has zero expectation, V_θ^X is almost surely equal to zero, and X is free. ∎

Theorem 2 *If* X *is sufficient,* $\mathscr{I}_X$ *is the information matrix associated with the given statistical space.*

Proof We can choose the measure which dominates the given statistical space, such that X is sufficient if and only if

$$p_\theta(\omega) = p_\theta^X(X(\omega))$$

and then if and only if

$$V_\theta(\omega) = V_\theta^X[X(\omega)].$$

The proof is completed by using Theorem A.1.1. ∎

Theorem 3 *Let X and Y be two independent statistics which admit $\mathscr{I}_X$ and $\mathscr{I}_Y$, respectively, as information matrices. Then the information matrix associated with (X, Y) exists and is equal to*

$$\mathscr{I}_{(X,Y)} = \mathscr{I}_X + \mathscr{I}_Y. \tag{1}$$

Proof Using obvious notation we have

$$p_\theta^{X,Y}(x, y) = p_\theta^X(x)p_\theta^Y(y)$$

and then

$$V_\theta^{X,Y}(X, Y) = V_\theta^X(X) + V_\theta^X(Y).$$

But, for every θ the statistics X and Y are independent. Thus the random vectors $V_\theta^X(X(\omega))$ and $V_\theta^Y(Y(\omega))$ are independent and have zero expectations implying (1). ∎

Corollary 1 *Let $(\Omega, \mathscr{A}; P_\theta, \theta \in \Theta)$ and $(\Omega', \mathscr{A}'; P'_\theta, \theta \in \Theta)$ be two dominated statistical spaces, having the same parameter and the same parameter space, which admit $\mathscr{I}$ and $\mathscr{I}'$, respectively, as information matrices. Then $\mathscr{I} + \mathscr{I}'$ is the information matrix of the weak product of these statistical spaces. In particular, $n\mathscr{I}$ is the information matrix of the sample $(\Omega, \mathscr{A}; P_\theta, \theta \in \Theta)^n$.*

Proof Apply Theorem 3 to the independent statistics

$$X:(\omega, \omega') \to \omega \qquad \text{and} \qquad Y:(\omega, \omega') \to \omega'.$$ ∎

Remark If we consider a product of statistical spaces, with obvious notation we have

$$\mathscr{I}(\theta, \theta') = \begin{pmatrix} \mathscr{I}(\theta) & 0 \\ 0 & \mathscr{I}'(\theta') \end{pmatrix}.$$

Definitions 1 and 2 and Jensen's inequality (Theorem A.4.9) imply

Theorem 4 *Let X be a statistic such that*

$$V_\theta^X(X) = E_{P_\theta}(V_\theta | X) \tag{2}$$

then

$$\mathscr{I}_X \leqslant \mathscr{I}.$$

On the other hand, the following argument shows that the condition considered in this theorem is rather general. By definition, we have

$$\forall C \in \mathscr{C}, \qquad P_\theta(X^{-1}(C)) = P_\theta^X(C),$$

that is,

$$\int_{X^{-1}(C)} p_\theta(\omega)\, d\nu = \int_C p_\theta^X(x)\, d\nu_X. \tag{3}$$

If we can differentiate (3) under the integral sign, we obtain

$$\forall C \in \mathscr{C}, \qquad \int_{X^{-1}(C)} V_\theta\, dP_\theta = \int_C V_\theta^X\, dP_\theta^X$$

and then (2), the condition of Theorem 4.

Remark 1 If the assumptions of Theorem 4 are satisfied, then

$$X = g(Y) \quad \Rightarrow \quad \mathscr{I}_X \leqslant \mathscr{I}_Y,$$

$$X \sim Y \quad \Rightarrow \quad \mathscr{I}_X = \mathscr{I}_Y.$$

Remark 2 Let X and Y be two statistics such that the conditional distribution of X given Y exists and admits a density $p_\theta^y(x)$ with respect to a measure which depends neither on y nor on θ. The following argument shows that we can consider the difference $\mathscr{I}_{(X,Y)} - \mathscr{I}_Y$ as a conditional information. Let

$$V_\theta^{X|Y}(\omega) = \operatorname{grad}_\theta \log p_\theta^{y(\omega)}(X(\omega)).$$

We have

$$p_\theta^{X,Y}(x, y) = p_\theta^Y(y) p_\theta^y(x)$$

and

$$V_\theta^{X,Y} = V_\theta^Y + V_\theta^{X|Y}.$$

But in general, the condition

$$\forall y, \qquad \int p_\theta^y(x)\, d\nu_X(x) = 1$$

implies that

$$E_{P_\theta}(V_\theta^{X|Y} \mid Y) = 0.$$

Then V_θ^Y and $V_\theta^{X|Y}$ are uncorrelated and finally

$$\mathscr{I}_{X|Y} = \mathscr{I}_{(X,Y)} - \mathscr{I}_Y = E_{P_\theta}[V_\theta^{X|Y} \cdot {}^t V_\theta^{X|Y}].$$

Theorem 5 (Cramer–Rao Inequality) *If $\mathscr{I}$ is nonsingular, if X is a vector valued statistic such that*

$$\Delta = E_{P_\theta}(X\,{}^t V_\theta) = \frac{d}{d\theta} E_{P_\theta}(X), \tag{4}$$

and if X has a covariance matrix Λ_X, then

$$\Lambda_X \geqslant \Delta \mathscr{I}^{-1}\,{}^t\Delta.$$

Proof The general term of the matrix Δ is equal to

$$\partial E_{P_\theta}(X_i)/d\theta_j, \qquad i = 1, \ldots, k; \quad j = 1, \ldots, s,$$

where $X = (X_1, \ldots, X_k)$ and $\theta = (\theta_1, \ldots, \theta_s)$. Condition (4) means that we can differentiate the following equality under the integral sign

$$E_{P_\theta}(X) = \int_\Omega X p_\theta(\omega)\, dv.$$

The theorem is proved if we show that the random vector

$$W = X - E_{P_\theta}(X) - \Delta \mathscr{I}^{-1} V_\theta$$

has the covariance matrix

$$\Delta_W = \Delta_X - \Delta \mathscr{I}^{-1}\,{}^t\Delta.$$

We have

$$\Lambda_W = E_{P_\theta}[(X - E_{P_\theta}(X) - \Delta \mathscr{I}^{-1} V_\theta)\,{}^t(X - E_{P_\theta}(X) - \Delta \mathscr{I}^{-1} V_\theta)]$$

or

$$\Lambda_W = \Lambda_X + \Delta \mathscr{I}^{-1} \Lambda_{V_\theta} \mathscr{I}^{-1}\,{}^t\Delta - E_{P_\theta}([X - E_{P_\theta}(X)]\,{}^t V_\theta) \mathscr{I}^{-1}\,{}^t\Delta$$
$$- \Delta \mathscr{I}^{-1} E_{P_\theta}(V_\theta \cdot {}^t(X - E_{P_\theta}(X))).$$

But, by definition, $\Lambda_{V_\theta} = \mathscr{I}$. On the other hand, we have $E_{P_\theta}(V_\theta) = 0$ and using (4), we obtain

$$\Lambda_W = \Lambda_X + \Delta \mathscr{I}^{-1}\,{}^t\Delta - \Delta \mathscr{I}^{-1}\,{}^t\Delta - \Delta \mathscr{I}^{-1}\,{}^t\Delta = \Lambda_X - \Delta \mathscr{I}^{-1}\,{}^t\Delta. \qquad \blacksquare$$

Remark If we consider $\Delta \mathscr{I}^{-1}\,{}^t\Delta$ as the lower boundary of dispersion of a statistic around its image, Corollary 1 shows that for a sample of size n, this boundary decreases to zero when n increases to infinity. On the other hand, we note that Condition (4) means that Δ can be expressed with only the help of the image of X.

Corollary 2 *If X, as given in Theorem 5, takes values in $(\mathbb{R}^s, \mathscr{B}_{\mathbb{R}^s})$ if Δ is nonsingular and if $\Lambda_X = \Delta \mathscr{I}^{-1}\,{}^t\Delta$, then there exist real functions $A(\theta)$, $K(\omega)$, and a function $H(\theta)$, which takes values in $\mathbb{R}^s$, such that*

$$\log p_\theta(\omega) = {}^tX(\omega) \cdot H(\theta) + A(\theta) + K(\omega). \tag{5}$$

Proof The covariance matrix, Λ_X is equal to zero, then W is almost surely equal to a constant. Therefore

$$V_\theta(\omega) = \mathscr{I}(\theta) \cdot \Delta^{-1}(\theta) \cdot [X(\omega) - E_{P_\theta}(X)].$$

Integration with respect to θ yields (5). ∎

Note that we obtain an exponential statistical space, which will be studied in Chapter 10.

Exercises

1. (a) Compute the information associated with the statistical space

$$(\mathbb{R}, \mathscr{B}_{\mathbb{R}}; N(m, \sigma^2), m \in \mathbb{R}, \sigma^2 \in \mathbb{R}^+)^n.$$

(b) Compute the information using σ, instead of σ^2, as the parameter.

2. Consider the following statistical space

$$\left(\mathbb{R}, \mathscr{B}_{\mathbb{R}}; \frac{\theta^a}{\Gamma(a)} e^{-\theta x} x^{a-1}, \theta \in \mathbb{R}^+\right)^n,$$

where a is a given positive number. Find the maximum likelihood statistic† and compute the information. Do the same problem if θ is known but a is unknown.

3. Compute the information in a sample from the Poisson distribution.

4. Consider a sample from the normal distribution, $N(m, \sigma^2)$, and three cases:

(a) m is known but σ unknown,

† The maximum likelihood statistic is defined in Section 6.2, Remark 2.

(b) m is unknown but σ known,

(c) both m and σ are unknown.

In each case, write the statistical space, compute the information and the variance of the maximum likelihood statistic. What interesting relationship exists between these two quantities?

5. Find conditions under which the converse of Theorem 2.2 is valid.

CHAPTER 4

Statistical Inference

This chapter is a little different from the others. Here we discuss the practical meaning of the mathematical tools used in statistics that have been developed and shall be developed in the rest of this book. To clarify the discussion which appears in every development of applied mathematics, we shall introduce some remarks to be differentiated from the usual mathematical definitions and theorems. Thus the reader can easily compare the mathematical concepts with their practical use, without confusing the mathematical arguments with the concrete motivations. We believe that this distinction will strengthen both mathematical and practical considerations.

1. Introduction

A statistical experiment is represented by a statistical space $(\Omega, \mathscr{A}; P_\theta, \theta \in \Theta)$. In previous chapters we studied the mathematical properties of these statistical spaces. Now we consider the use of the statistic formed from the data, i.e., we shall discuss *statistical inference*.

In every statistical problem we have data represented by ω, and want to draw some conclusion about the unknown parameter θ. This conclusion, for example, may be to make a decision or to represent the data by a distribution. *Testing statistical hypotheses* and *estimation* are classical examples of such problems. The following general model of statistical decision admits, as special cases, many of the usual statistical problems. However, we must note that these problems can be solved without the help of this general model (see Chapters 5 and 6). Moreover, some problems of data analysis do not involve decision models. The reader will note that a statistical argument in a practical problem is an inductive one. Thus he must give more of his attention to the practical meaning of the concepts than to the mathematical development. The theory of statistical decision is developed in Wald [63] and one can find in Blackwell and Girschik [8] and Ferguson [18] connections between statistics and game theory (see also [24, 25]).

We should recall that the solution deduced from a mathematical theory is not to be considered decisive for a practical problem. There are two reasons for this special feature of statistical inference. First, the hypothesis on which the model (the statistical space) is built is always arbitrary; second, the observations are basically made through a random process.

2. Decisions and Strategies

Definition 1 A *statistical decision problem* is defined by a statistical space $[\Omega, \mathscr{A}; P_\theta, \theta \in (\Theta, \mathscr{T})]$ and a measurable space $(\Delta, \mathscr{D})$. A transition probability function $S_\omega(D)$ on $\Omega \times \mathscr{D}$ will be called a *strategy* S.

Remark 1 The space $(\Delta, \mathscr{D})$ is called a ***decision space***. This is because we assume that the information requested on θ is the information necessary to make a decision among the family Δ of decisions, which are given in advance. When the observation ω is given, a decision δ is taken by using the distribution S_ω on $(\Delta, \mathscr{D})$. Specifically if for every ω, S_ω is concentrated at the point $s(\omega) \in \Delta$, the corresponding strategy is called ***deterministic***; it consists in taking decision $s(\omega)$ when the observation ω is given. For example, see Exercise 4.

Remark 2 Mathematically, it can be seen why we would not want to limit our strategies to deterministic ones. The set of all strategies is convex. On the other hand, the subset of all deterministic strategies is not convex. From a practical point of view, a nondeterministic strategy can appear somewhat unrealistic. However, we observe that the distribution S_ω, defined for every ω when the strategy S is given, can be termed a *preference* function on all possible decisions.

Remark 3 The field $\mathscr{D}$ has essentially a mathematical use. In some cases (e.g., set estimation) it is difficult to choose such a subfield. If this subfield cannot be chosen, theoretically, the problem being considered is not a statistical decision problem.

Remark 4 In many cases a subset C^* of $\Theta \times \Delta$ and belonging to $\mathscr{T} \otimes \mathscr{D}$ is given. From a practical viewpoint this is the set of "correct" decisions; i.e., if θ is the true value of the parameter, the decision δ which is taken is consistent with θ if (θ, δ) belongs to C^*. For example, find the subset C^* relative to Exercise 4.

Definition 2 The *image* of strategy S is the transition probability function $\mathscr{I}^S$ which is the transposition product of P_θ and S,

$$\forall C \in \mathscr{D}, \qquad \mathscr{I}_\theta^S(C) = \int_\Omega S_\omega(C)\, dP_\theta(\omega).$$

In particular, if the strategy S is the deterministic strategy $s(\omega)$, then we have

$$\forall C \in \mathscr{D}, \qquad \mathscr{I}_\theta^s(C) = \int_{s^{-1}(C)} dP_\theta(\omega).$$

The image of s as a strategy is for every θ the distribution of the statistic s taking values in $(\Delta, \mathscr{D})$. Note that the image of s as a statistic is not the image of s as a strategy. See Definition 1.2.8 and Remark 3.1 below.

Remark 5 In mathematical statistics, the study of a strategy is made through its image. This means that the practical value of a strategy is defined by its image.

Theorem 1 *Let $\mathscr{B}$ be a sufficient subfield for the statistical space $(\Omega, \mathscr{A};$ $P_\theta, \theta \in (\Theta, \mathscr{T}))$, let $(\Delta, \mathscr{D})$ be a decision space and let S be a strategy. Every strategy Σ, in which for every $C \in \mathscr{D}$, $\Sigma_\omega(C)$ is a determination of $E[S_\omega(C) | \mathscr{B}]$ is called the* projection *of S on $\mathscr{B}$. The strategies S and Σ have the same image.*

Proof For every $C \in \mathscr{D}$, the probability $S_\omega(C)$ considered as a function of ω is a statistic which has a projection on $\mathscr{B}$. The definition of Σ means that for every C we can choose a determination of $E[S_\omega(C) | \mathscr{B}]$ which, for every given ω, is a distribution on $(\Delta, \mathscr{D})$. Suppose, for example, that the given statistical space is dominated by a special distribution P^*. Let π be the distribution induced on $(\Omega \times \Delta, \mathscr{A} \otimes \mathscr{D})$ by (P^*, S). If there exists a regular determination of the conditional probability of π given $\mathscr{B} \otimes \Delta$, then the previous condition is fulfilled; i.e., for every $C \in \mathscr{D}$ we can choose

$$\Sigma_\omega(C) = \pi(\Omega \times C | \mathscr{B} \otimes \Delta).$$

Finally, we have

$$\forall C \in \mathscr{D}, \qquad \mathscr{I}_\theta^S(C) = E_{P_\theta}(S_\omega(C)) = E_{P_\theta}[E(S_\omega(C) | \mathscr{B})] = E_{P_\theta}[\Sigma_\omega(C)] = \mathscr{I}_\theta^\Sigma(C)$$

and Σ and S have the same image. ∎

Remark 6 If Σ exists, we see that $\Sigma_\omega(C)$ is a $\mathscr{B}$-measurable function of ω and then Σ is a strategy defined on

$$(\Omega, \mathscr{B}; P_\theta, \theta \in (\Theta, \mathscr{T})). \tag{1}$$

The Theorem shows that if a sufficient subfield $\mathscr{B}$ exists, we can, under general conditions, use the statistical space (1), instead of the given statistical space. Specifically, if the subfield $\mathscr{B}$ is induced by a sufficient statistic X, we substitute the statistical space induced by X for the initial statistical space. Note that the projection of a deterministic strategy is not, in general, a deterministic strategy.

3. Hypothesis Testing and Statistical Estimation

Testing statistical hypotheses and estimation are the most classic examples of statistical decisions. We shall distinguish carefully between a test and a test of one hypothesis against another.

Definition 1 Let $(\Omega, \mathscr{A}, \mathscr{P})$ be a statistical space. A nonempty subset of $\mathscr{P}$ is called a *hypothesis* and a statistic taking values in $[0, 1]$ is called a *test*. If $\mathscr{P} = \{P_\theta, \theta \in \Theta\}$ we also call a nonempty subset of Θ a *hypothesis.*

Definition 2 Let $(\Omega, \mathscr{A}, \mathscr{P})$ be a statistical space and let $\mathscr{P}_0$ and $\mathscr{P}_1$ be two disjoint hypotheses. We call a *test of* $\mathscr{P}_0$ *against* $\mathscr{P}_1$ a test Φ to which the following strategy is associated. For every observation ω the two hypotheses $\mathscr{P}_0$ and $\mathscr{P}_1$ take probabilities $1 - \Phi(\omega)$ and $\Phi(\omega)$, respectively. The image β_Φ of the test Φ, considered as a function on $\mathscr{P}_0 \cup \mathscr{P}_1$ only, is called the *power function of* Φ.

Remark 1 This definition is a special case of a statistical decision, defined previously. The decision space Δ has two elements, $\{0\}$ and $\{1\}$, corresponding to the choice of $\mathscr{P}_0$ or $\mathscr{P}_1$ respectively. The subfield $\mathscr{D}$ is composed of the four subsets of Δ. The pair $[1 - \Phi(\omega), \Phi(\omega)]$ defines a transition probability function on $\Omega \times \mathscr{D}$. If furthermore, a parameter θ is given and

$$\mathscr{P} = \{P_\theta, \theta \in (\Theta, \mathscr{T})\},$$

where P_θ is a transition probability function, the pair $[1 - \beta_\Phi(\theta), \beta_\Phi(\theta)]$ defines on $\Theta \times \mathscr{D}$ a transition probability function which is the image of the strategy associated with Φ. Note that in Definition 2, Φ is considered as a statistic. Its image is given in Definition 1.2.8 and it should not be confused with the image of a strategy. Finally, the set of correct decisions is defined by

$$C^* = \{\Theta_0 \times 0\} \cup \{\Theta_1 \times 1\}.$$

Definition 3 Let $(\Omega, \mathscr{A}; P_\theta, \theta \in (\Theta, \mathscr{T}))$ be a statistical space and let f be a measurable mapping from $(\Theta, \mathscr{T})$ to $(\mathscr{X}, \mathscr{C})$. Every statistic taking values in $(\mathscr{X}, \mathscr{C})$ is called an *estimator* of f.

Definition 4 A mapping d from Ω to $\mathscr{C}$ is called the *set estimator* of f if

$$\forall x \in \mathscr{X}, \qquad d^{-1}(x) = \{\omega : x \in d(\omega)\} \in \mathscr{A}.$$

For a given ω, $d(\omega)$ is called a *confidence region* for f.

Remark 2 Estimation and set estimation are complementary as in physics where we associate a measurement with a bound of the error. Here an estimator g gives a representative value of $f(\theta)$ and the confidence region $d(\omega)$ localizes $f(\theta)$. Frequently, $g(\omega)$ is a central value of $d(\omega)$.

Remark 3 Taking $(\mathscr{X}, \mathscr{C})$ as a decision space yields an estimator which is a deterministic strategy. The set C^* corresponding to correct decisions is the surface on $\mathscr{X} \times \Theta$ defined by $f(\theta) = x$. Generally, a strategy is a transition probability on $\Omega \times \mathscr{C}$, but Theorem 5.1 will show that in most cases, one can consider estimators only. On the contrary, in testing hypotheses, one cannot restrict oneself to deterministic strategies. Moreover, we shall see in Chapters 5 and 6 that we cannot estimate a parameter by testing each possible value and, conversely, that we cannot test two hypotheses by estimating their indicator functions.

Remark 4 We shall not consider set estimation as a statistical decision problem since we cannot choose a subfield of $\mathscr{C}$ for which d becomes a measurable mapping. We consider set estimation as resulting first from the choice of a strategy S which for every ω gives a distribution S_ω on $(\mathscr{X}, \mathscr{C})$. Then, with respect to this distribution, we can choose a set $d(\omega)$ having some optimal property.

Definition 5 A subset D of $\Omega \times \mathscr{X}$, belonging to $\mathscr{A} \otimes \mathscr{C}$ is called a *strong set estimator* (or a set estimator having a measurable graph) of f.

We see easily that for each strong set estimator we can deduce a set estimator; for every ω, $d(\omega)$ is the section of D with respect to ω and $d^{-1}(\theta)$ is the section of D with respect to $f(\theta)$. But conversely, let d be a set estimator, then the locus which is defined in $\Omega \times \mathscr{X}$ by the pairs $(\omega, d(\omega))$ where ω traverses Ω, is not necessarily a strong set estimator.

4. Choosing a Strategy

Remark 1 A fundamental statistical-decision problem is the choice of a strategy with optimal property. Mathematically it is natural to introduce a quasi ordering on strategies, and we note that the following definitions are those of *maximum* element, *maximal* element, and *cofinal* set, respectively.

Definition 1 If a quasi-ordering on the strategies has been chosen we say that the strategy S is *optimal* if it is better than or as good as any other strategy and that the strategy S is *admissible* if a better one does not exist.

Definition 2 We say that the family $\mathscr{S}$ is *complete* with respect to a given quasi-ordering if for every strategy S there exists a strategy S' belonging to $\mathscr{S}$ such that S' is better than or as good as S.

Remark 2 Choosing a quasi-ordering is always important and difficult. Suppose that a set C^* of correct decisions has been given. The strategy S is better than S' if

$$\forall \theta \in \Theta, \qquad \begin{cases} \mathscr{I}_\theta^S(C_\theta^*) \geqslant \mathscr{I}_\theta^{S'}(C_\theta^*), \\ \mathscr{I}_\theta^S(\complement C_\theta^*) \leqslant \mathscr{I}_\theta^{S'}(\complement C_\theta^*), \end{cases}$$

where C_θ^* is the section of C^* corresponding to θ. However, this quasi ordering is too strong and in many cases (e.g., hypothesis testing) an optimal strategy does not exist.

The author believes that in statistics, no general or reasonable method of choosing a strategy exists as there is no realistic quasi-ordering for which we can effectively compute an admissible strategy. In this connection we use a quasi-ordering specifically to eliminate all strategies which are not admissible. We shall select a strategy adapted to each particular case. The viewpoint for estimation is different from that of hypothesis testing. Even, in hypothesis testing there exist great differences among the elementary parametric case (see Chapter 11), the more complicated parametric cases, such as the Behrens–Fisher problem (see Section 10.5) and the nonparametric case for which we can only compute the level of significance for some well-known tests.

5. Quasi-Ordering Induced by a Loss Function

Definition 1 Using notation of Definition 2.1 a measurable mapping W, such that

$$W:(\Theta, \mathscr{T}) \times (\Delta, \mathscr{D}) \rightarrow (\bar{\mathbb{R}}^+, \mathscr{B}_{\bar{\mathbb{R}}^+}),$$

is called a *loss function*.

Remark 1 In practice $W(\theta, \delta)$ represents the loss corresponding to the decision δ when the true value is θ. We assume that W is positive without loss of generality and thereby simplify some proofs. Moreover, if a set C^* of correct decisions is given, in many cases W is equal to zero on C^* and it is reasonable to assume that

$$\sup[W(\theta, \delta) \mid (\theta, \delta) \in C^*] \leqslant \inf[W(\theta, \delta) \mid (\theta, \delta) \in \complement C^*].$$

Remark 2 The notion of loss function is enticing but in many cases somewhat arbitrary. It implies that the loss associated with C^* and with $\complement C^*$ can be measured with respect to the same monetary unit. See Exercise 4 for an example. If the decisions to be made have human or social consequences, this implication is usually not a reasonable one.

Definition 2 Let S be a strategy and W a loss function. We define
(a) $W_S(\theta, \omega)$, the *expected loss* as the function on $\Theta \times \Omega$,

$$W_S(\theta, \omega) = \int_\Delta W(\theta, \delta)\, dS_\omega(\delta);$$

(b) $R_S(\theta)$, the *risk* as the function on Θ,

$$R_S(\theta) = \int_\Omega W_S(\theta, \omega)\, dP_\theta(\omega) = \int_\Delta W(\theta, \delta)\, d\mathscr{I}_\theta^S(\delta).$$

Moreover, if a prior distribution Q is given, we define R_S^Q the *average risk*, as

$$R_S^Q = \int_\Theta R_S(\theta)\, dQ(\theta) = \iint_{\Theta \times \Delta} W(\theta, \delta)\, d(Q(\theta) \cdot \mathscr{I}_\theta^S(\delta)).$$

Note that since W is positive, all integrals in this definition are defined and either finite or infinite. The above equalities result from Theorem A.6.1 applied to $(Q, \mathscr{I}_\theta^S)$ and applied, for every θ, to (P_θ, S).

In the case of a vector-valued estimator, we use a vector-valued loss function.

Definition 3 Using notation of Definition 3.3, let $\mathscr{V}$ be a finite dimensional linear space with a linear ordering and $\mathscr{V}^+$ be the positive cone of $\mathscr{V}$. Then W, the *loss function*, is defined as the mapping

$$W: \Theta \times \mathscr{X} \to \mathscr{V}^+.$$

Remark 3 If we choose x as the value of $f(\theta)$ and if θ is the true value of the parameter, then the corresponding loss is represented by $W(\theta, x)$. If, for example, $\mathscr{X}$ is a normed linear space, we often choose

$$W(\theta, x) = a(\theta) \|x - f(\theta)\|.$$

But, if $\mathscr{X} = \mathbb{R}^k$, we take for $\mathscr{V}$ the space of symmetric matrices of order k, for $\mathscr{V}^+$ the cone of the symmetric matrices of order k which are nonnegative definite and for loss function,

$$W(\theta, x) = [x - f(\theta)]^t[x - f(\theta)].$$

Definition 4 Let W be a loss function. The corresponding risk $R_S(\theta)$ associated with a given strategy S defines a *quasi-ordering* by

$$S \overset{W}{\geqslant} S' \quad \Leftrightarrow \quad \{\forall \theta \in \Theta, \quad R_S(\theta) \leqslant R_{S'}(\theta)\},$$

and by $S \overset{W}{>} S'$ if one of the above inequalities is strict.

Definition 5 A class $\mathscr{W}$ of loss functions defines a *quasi-ordering on strategies* by

$$S \overset{\mathscr{W}}{\geqslant} S' \quad \Leftrightarrow \quad \{\forall W \in \mathscr{W}, \quad S \overset{W}{\geqslant} S'\},$$

and by $S \overset{\mathscr{W}}{>} S'$ if for one loss function $W \in \mathscr{W}$ we have $S \overset{W}{>} S'$.

Remark 4 Definition 5 is as useful as Definition 4, because, in some cases, a class of loss functions can be chosen more easily than a loss function alone (see, e.g., Section 5.2).

Definition 6 A loss function W and a prior distribution Q define a quasi-ordering (called *Bayesian*) on strategies by

$$S \underset{Q}{\overset{W}{\geqslant}} S' \quad \Leftrightarrow \quad R_S^Q \leqslant R_{S'}^Q.$$

Note that if W belongs to $\mathscr{W}$ we have

$$S \overset{\mathscr{W}}{\geqslant} S' \quad \Rightarrow \quad S \overset{W}{\geqslant} S' \quad \Rightarrow \quad S \underset{Q}{\overset{W}{\geqslant}} S', \qquad \forall Q.$$

Remark 5 Other quasi-orderings on strategies exist. For example, if μ is a positive measure on $(\Theta, \mathscr{T})$, we can weaken Definitions 4 and 5 by defining the quasi-ordering by

$$R_S(\theta) \leqslant R_{S'}(\theta), \qquad \mu\text{-a.e. on } \Theta.$$

Theorem 1 *Suppose that Δ is a convex and bounded Borel subset of $\mathbb{R}^k$. Let $\mathscr{W}_0$ be a family of loss functions $W(\theta, \delta)$ which for all $\theta \in \Theta$ are continuous and convex as function of δ. Then the family of deterministic strategies is complete with respect to the quasi-ordering defined by $\mathscr{W}_0$.*

Proof Since Δ is a convex and bounded Borel set, the following deterministic strategy is well defined, for every strategy S, by Theorem A.6.1:

$$s(\omega) = \int_\Delta \delta \, dS_\omega(\delta).$$

On the other hand, applying Jensen's inequality for every ω to the distribution S_ω on $(\Delta, \mathscr{D})$, we have

$$W(\theta, s(\omega)) \leqslant \int_\Delta W(\theta, \delta) \, dS_\omega(\delta).$$

Then, for every fixed θ, by integrating with respect to $P_\theta(\omega)$ we have

$$R_S(\theta) \geqslant R_s(\theta), \qquad \forall\theta \in \Theta. \qquad \blacksquare$$

This theorem can be used in estimation theory (see Section 6.2), but not for hypothesis testing. Moreover, we can weaken the condition that Δ be bounded (see Ferguson [18, p. 76]). However, under general conditions (see Ferguson [18, Chapter 2]) the following propositions are true.

(a) For every pair (W, Q) there exists an optimal strategy with respect to the quasi-ordering defined by this pair.

(b) If the strategy S is optimal with respect to the quasi-ordering defined by (W, Q), then S is Q-almost surely admissible with respect to the quasi-ordering defined by W.

(c) If S is admissible with respect to the quasi-ordering defined by W, there exists a prior distribution Q for which S is optimal with respect to the quasi-ordering defined by (W, Q).

(d) The class of strategies which are optimal with respect to the quasi-ordering defined by (W, Q), when Q runs over all possible distributions, is a complete class with respect to the quasi-ordering defined by W.

6. Nuisance Parameters

Definition 1 Let $(\Omega, \mathscr{A}; P_\theta, \theta = (\lambda, \mu) \in \Theta = \Lambda \times M)$ be a statistical space, Δ a decision space and C^* a subset of $\Delta \times \Theta$. If there exists a subset C_0^* of $\Delta \times \Lambda$ such that

$$C^* = C_0^* \times M,$$

we say that μ is a *nuisance* parameter with respect to C^*.

Sometimes we say that λ is the *principal* parameter.

Remark 1 Note that the subfield $\mathscr{D}$ of the decision space does not appear in this definition. Thus, we can apply the definition to set estimation. A nuisance parameter is an unknown parameter on which we do not seek information. The concept of nuisance parameter is defined with respect to a given specific problem in statistics since C^* will always be the subset of correct decisions. In many classical problems of estimation or hypothesis testing there exist nuisance parameters (see Linnik [41]), which always complicate the problem.

Example 1 If we test Θ_0 against Θ_1, where Θ_0 and Θ_1 are disjoint subsets of $\Lambda \times M$, there exists a nuisance parameter if

$$\Theta_0 = \Lambda_0 \times M, \qquad \Theta_1 = \Lambda_1 \times M, \qquad \Lambda_0 \subset \Lambda, \quad \Lambda_1 \subset \Lambda.$$

Example 2 If we estimate a function f which is not a one-to-one mapping, we generally have a nuisance parameter.

Remark 2 Frequently, we shall have to change the variables and the parameters to fit the situation described in Definition 1.

Definition 2 Let $(\Omega, \mathscr{A}; P_\theta, \theta = (\lambda, \mu) \in \Lambda \times M)$ be a statistical space. We say that the subfield $\mathscr{B}$ is sufficient (free, complete or quasi-complete), with respect to the parameter λ, if it is sufficient (free, complete or quasi-complete), for every fixed value of μ. We say that a statistic is distribution-free (or mean-free) with respect to the parameter λ if its distribution (or mean) does not depend on λ.

We can see easily that if the subfield $\mathscr{B}$ is sufficient or free, it is sufficient or free with respect to any component of the parameter.

To eliminate nuisance parameters, the following procedure is often useful (see Section 11.4 and Exercises 11.1, 11.5, 11.15). Suppose that

(a) $(\Omega, \mathscr{A}) = (\mathscr{X} \times \mathscr{Y}, \mathscr{C} \otimes \mathscr{C}')$,

(b) For every θ, there exists a regular determination of the conditional probability of X given Y, where X and Y are the two projections $(x, y) \to x$ and $(x, y) \to y$.

(c) Y is sufficient with respect to μ.

Then the conditional distribution of X given Y does not depend on μ. Let P_λ^y be this distribution. Furthermore, the statistical space

$$(\mathscr{X}, \mathscr{C}; P_\lambda^y, \lambda \in \Lambda)$$

does not depend on the nuisance parameter but on the conditioning variable y and thus we have substituted a known variable for an unknown parameter. Let S^y be a strategy on this statistical space, i.e.,

$$S^y : (x, D) \to S_x^y(D).$$

The image $\mathscr{I}_\lambda^y$ of S^y does not depend on μ. If the mapping

$$S:(x, y; D) \to S_x^y(D)$$

defines a strategy S on the given statistical space (this is a general condition of measurability), then the image of S is given by

$$\mathscr{I}_\theta^S = \int_{\mathscr{Y}} \mathscr{I}_\lambda^y \, dP_Y^\theta(y),$$

where P_Y^θ is the distribution of Y.

Exercises

1. Show that if Δ is a countable set, a strategy taking values on $(\Delta, \mathscr{D})$ is equivalent to a deterministic strategy taking values on the family of all distributions on $(\Delta, \mathscr{D})$.

2. Show that if $\Delta = \mathbb{R}^k$ and if $s(\omega)$ is an integrable deterministic strategy, the projection on $\mathscr{B}$ of the statistic s is the expectation of the projection on $\mathscr{B}$ of the strategy s.

3. Using the terminology of statistical decisions, generalize hypothesis testing to the case where we must choose among a finite number, greater than two, of hypotheses. What is the image of a strategy?

4. Suppose that for the quality control of a homogeneous production lot we observe the number of defectives in a sample of size n. Show that this problem can be considered as a statistical decision problem and suggest a strategy. Give conditions under which a loss function can be defined.

5. Consider the following statistical space $[\mathbb{R}, \mathscr{B}_{\mathbb{R}}; N(m, 1), m \in \mathbb{R}]^n$ and let m_0, m_1, m_2 be three given numbers such that

$$m_0 < m_1 < m_2 .$$

We want to choose among the three hypotheses

$$\{m = m_0\}, \qquad \{m = m_1\}, \qquad \{m = m_2\}.$$

Let c and d ($c < d$) be two given numbers and let $\bar{x}$ be the sample mean. Consider the following strategy,

$$\begin{array}{ll} \text{if } \bar{x} \leqslant c & \text{we choose the hypothesis } \{m = m_0\}; \\ \text{if } c < \bar{x} \leqslant d & \text{we choose the hypothesis } \{m = m_1\}; \\ \text{if } d < \bar{x} & \text{we choose the hypothesis } \{m = m_2\}. \end{array}$$

(a) How many types of errors can occur? Compute the probability of these errors as functions of c and d.

(b) Find the values of c and d which correspond to the strategy where we choose the maximum likelihood hypothesis. Compute c, d, and the probabilities of errors in the special case $n = 25$, $m_0 = -1$, $m_1 = 0$, $m_2 = 1$.

(c) Suppose that $m_0 = -m_2$, $m_1 = 0$. Show that the values of c and d for which the maximum of the error probabilities is minimum are

$$c = -x, \qquad d = x,$$

where x is the unique root of an equation to be found.

CHAPTER 5

Testing Statistical Hypotheses

The testing of statistical hypotheses was introduced in Section 4.3. In this chapter we develop the theory of nonsequential tests. Assimilation of these ideas cannot be made without numerous simple exercises, such as are given here. Further applications appear in Chapters 9 and 11.

1. Definitions and Preliminary Remarks

Definition 1 The tests Φ and Φ' of $\mathscr{P}_0$ against $\mathscr{P}_1$ are *equivalent* if they have the same power function.

This definition must not be confused with the definition of equivalence of a test when it is considered as a statistic. Here the definition means that the statistical value (from a practical viewpoint) of a test as a decision function between two hypotheses is given entirely by its power function. Note that if we test one hypothesis against another, we can make two wrong decisions: choosing $\mathscr{P}_1$ when P belongs to $\mathscr{P}_0$ or, conversely, choosing $\mathscr{P}_0$ when P belongs to $\mathscr{P}_1$. In the first case, the value of the test Φ is deduced from its power function on $\mathscr{P}_0$ and in the second case, from its power function on $\mathscr{P}_1$.

Definition 2 Let Φ be a test of $\mathscr{P}_0$ against $\mathscr{P}_1$. We call α_Φ the *level of significance* of Φ, where

$$\alpha_\Phi = \sup(\beta_\Phi(P) \mid P \in \mathscr{P}_0).$$

We say that Φ is *unbiased* if

$$\alpha_\Phi \leqslant \beta_\Phi(P), \qquad \forall P \in \mathscr{P}_1.$$

Definition 3 The test Φ of $\mathscr{P}_0$ against $\mathscr{P}_1$ is *deterministic* if the event belonging to $\mathscr{A}$ and defined by

$$\Phi(\omega)[1 - \Phi(\omega)] > 0$$

is $\mathscr{P}_0 \cup \mathscr{P}_1$-negligible. In this case we call the event

$$\{\omega : \Phi(\omega) = 1\}$$

the *critical region* of Φ.

If a test is not deterministic we say that it is *randomized.*

Definition 4 A test Φ of $\mathscr{P}_0$ against $\mathscr{P}_1$ is *trivial* (*or powerless*) if it is a constant function on Ω (or if its power function is a constant).

Note that the previous definitions depend on the order in which the two hypotheses are given. If we permute $\mathscr{P}_0$ and $\mathscr{P}_1$ we must change Φ to $1 - \Phi$.

Theorem 1 *Let U be the uniform distribution on $[0, 1]$ and let $\mathscr{B}$ be the Borel subfield of this interval. If Φ is a test on the statistical space $(\Omega, \mathscr{A}, \mathscr{P})$, the deterministic test defined on the product statistical space,*

$$(\Omega, \mathscr{A}, \mathscr{P}) \otimes ([0, 1], \mathscr{B}, U)$$

by the critical region

$$C = \{(\omega, u) \in \Omega \times [0, 1] : \Phi(\omega) > u\}$$

has the same image as Φ.

Proof The indicator function $1_C(\omega, u)$ of the set C is $\mathscr{A} \otimes \mathscr{B}$-measurable. For every P in $\mathscr{P}$, let

$$P' = P \times U.$$

Using Fubini's theorem we have

$$\beta_{1_C}(P') = P'(C) = \int_\Omega \left[\int_0^1 1_C(\omega, u)\, dU(u) \right] dP = \int_\Omega \Phi(\omega)\, dP = \beta_\Phi(P). \quad \blacksquare$$

This theorem shows that by using a random variable (having the distribution U) which is independent of the given problem, we can transform any test to an equivalent deterministic test.

The following theorem follows directly from Definitions 1 and 2.3.1 and shows that if we have a sufficient statistic we can restrict ourselves to tests which are functions of this statistic only.

Theorem 2 *Let Φ be a test of $\mathscr{P}_0$ against $\mathscr{P}_1$. If $\mathscr{B}$ is a sufficient subfield on $(\Omega, \mathscr{A}; \mathscr{P}_0 \cup \mathscr{P}_1)$, $E(\Phi | \mathscr{B})$, the projection of Φ on $\mathscr{B}$ defines a test which is equivalent to Φ.*

Remark The projection on a sufficient subfield of a deterministic test is, in general, a randomized test. Conversely, if we have a randomized test which is a function of a sufficient statistic, it is sometimes useful to seek a deterministic test which is equivalent to the given test but not necessarily a function of the sufficient statistic only. This operation is especially interesting if we can replace a given test which is irregular or complicated by an equivalent but simpler test which does not depend *only* on a sufficient statistic. From a mathematical viewpoint, if a $\mathscr{B}$-measurable test Φ is given, we must find an event A belonging to $\mathscr{A}$ such that

$$P(A | \mathscr{B}) = \Phi, \qquad \forall P \in \mathscr{P}. \tag{1}$$

In particular, if the statistical space is dominated by the special distribution P^*, (1) is equivalent to

$$P^*(A | \mathscr{B}) = \Phi.$$

We dealt with this type of problem in Section 2.7. To obtain the solution, first perform a transformation of variables such that the sufficient statistic x and a free statistic y appear separately. Then for every value of x find an event A_x concerning y, having probability $\Phi(x)$ such that the event A, whose sections are A_x, is measurable.

Definition 5 The family of power functions of all possible tests of one given hypothesis against another is called the *space associated* with this given hypothesis-testing problem.

Remark Theorem 12.3.5 will show that if the statistical space is dominated, then the associated space is compact for the weak topology $\sigma\{\mathscr{L}_\infty, \mathscr{M}\}$, defined in Section 12.3.

Definition 6 A test Φ of $\mathscr{P}_0$ against $\mathscr{P}_1$ is *free* if its power function is constant on $\mathscr{P}_0$, i.e., if

$$\beta_\Phi(P) = \alpha_\Phi, \qquad \forall P \in \mathscr{P}_0.$$

This means that the *statistic* Φ defined on $(\Omega, \mathscr{A}, \mathscr{P}_0)$ is *mean-free.*

Theorem 3 *Let Θ_0, Θ_1 be two disjoint hypotheses on the statistical space*

$$[\Omega, \mathscr{A}; P_\theta, \theta \in (\Theta, \mathscr{T})].$$

If a topology on Θ is given such that every test Φ has a continuous image, then for every unbiased test Φ of Θ_0 against Θ_1 we have

$$\forall \theta \in \bar{\Theta}_0 \cap \bar{\Theta}_1, \qquad \beta_\Phi(\theta) = \alpha_\Phi.\dagger$$

Proof If Φ is an unbiased test of Θ_0 against Θ_1 we have

$$\beta_\Phi(\theta) \leqslant \alpha_\Phi, \qquad \forall \theta \in \Theta_0; \qquad \beta_\Phi(\theta) \geqslant \alpha_\Phi, \qquad \forall \theta \in \Theta_1.$$

Since β_Φ is continuous, these inequalities remain valid on $\bar{\Theta}_0 \cap \bar{\Theta}_1$. ∎

Definition 7 Let $(\Omega, \mathscr{A}, \mathscr{P})$ be a statistical space and let $\mathscr{B}$ be a subfield of $\mathscr{A}$. A test Φ of $\mathscr{P}_0$ against $\mathscr{P}_1$ is *free given* $\mathscr{B}$, if the statistic Φ defined on the statistical space $(\Omega, \mathscr{A}, \mathscr{P}_0)$ is mean-free given $\mathscr{B}$.

Usually $\mathscr{B}$ is a sufficient statistic and then we say that Φ has *Neyman structure* on the statistical space $(\Omega, \mathscr{B}, \mathscr{P}_0)$. In this case, if the statistical space $(\Omega, \mathscr{B}, \mathscr{P}_0)$ is quasi-complete, Theorem 2.3.3. implies that every free test of $\mathscr{P}_0$ against $\mathscr{P}_1$ has Neyman structure on the statistical space $(\Omega, \mathscr{B}, \mathscr{P}_0)$.

2. Quasi-Ordering on Tests of Hypotheses

The basic problem of hypothesis testing is to find the best test of one hypothesis against another. It will be convenient to begin our study with quasi-ordering on tests.

Definition 1 Let Φ and Φ' be two tests of $\mathscr{P}_0$ against $\mathscr{P}_1$. We write

$$\Phi \geqslant \Phi' \quad \Leftrightarrow \quad \begin{cases} \beta_\Phi(P) \leqslant \beta_{\Phi'}(P), & \forall P \in \mathscr{P}_0, \\ \beta_\Phi(P) \geqslant \beta_{\Phi'}(P), & \forall P \in \mathscr{P}_1, \end{cases}$$

and $\Phi > \Phi'$ if, in addition, one of the above inequalities is strict for a distribution $P \in \mathscr{P}_0 \cup \mathscr{P}_1$.

The quasi-ordering is indispensable but not sufficient to determine admissible tests. If we are given a loss function, or a loss function and a prior distribution, we can introduce the quasi-ordering on the strategies which are associated with tests of one hypothesis against another, by using the definitions of Section 4.5. Then consider the statistical space $[\Omega, \mathscr{A}; \mathrm{P}_\theta, \theta \in (\Theta, \mathscr{T})]$, a loss function W and two disjoint hypotheses Θ_0 and Θ_1.

† $\bar{\Theta}_0$ is the smallest *closed* set containing Θ_0. It is usually called the *closure*. (Ed.)

Using the terminology of Remark 4.3.1, the function W whose range of values is $(\overline{\mathbb{R}}^+, \mathscr{B}_{\overline{\mathbb{R}}^+})$, is defined on $(\Theta_0 \cup \Theta_1) \times \Delta$ and is $\mathscr{T} \otimes \mathscr{D}$-measurable.

The risk R_Φ corresponding to a test Φ of Θ_0 against Θ_1 is

$$\begin{aligned} R_\Phi(\theta) &= E_{P_\theta}[W(\theta,0)[1-\Phi(\omega)] + W(\theta,1)\Phi(\omega)] \\ &= W(\theta,0) + [W(\theta,1) - W(\theta,0)]\beta_\Phi(\theta). \end{aligned} \tag{1}$$

Theorem 1 *Let $\mathscr{W}$ (or $\mathscr{W}'$) be the family of loss functions W such that*

$$W(\theta,1) \leqslant W(\theta,0) \qquad (\text{or } W(\theta,1) < W(\theta,0)), \qquad \forall \theta \in \Theta_1$$

$$W(\theta,1) \geqslant W(\theta,0) \qquad (\text{or } W(\theta,1) > W(\theta,0)), \qquad \forall \theta \in \Theta_0.$$

Then we have

$$\Phi \geqslant \Phi' \;\Rightarrow\; \Phi \overset{\mathscr{W}}{\geqslant} \Phi'; \qquad \Phi > \Phi' \;\Rightarrow\; \Phi \overset{\mathscr{W}'}{>} \Phi';$$

$$\forall W \in \mathscr{W}', \qquad \Phi \overset{W}{\geqslant} \Phi' \;\Rightarrow\; \Phi \geqslant \Phi'; \qquad \Phi \overset{W}{>} \Phi' \;\Rightarrow\; \Phi > \Phi'.$$

This theorem follows directly from (1) and Definition 1. It shows that the quasi-ordering which is defined by Definition 1 is fundamental for hypothesis testing.

Theorem 2 *Let W be a loss function belonging to the family $\mathscr{W}$ and let Q be a prior distribution on $\Theta_0 \cup \Theta_1$ such that the integrals*

$$A = \int_{\Theta_0} [W(\theta,1) - W(\theta,0)]\, dQ(\theta), \qquad B = \int_{\Theta_1} [W(\theta,0) - W(\theta,1)]\, dQ(\theta)$$

are finite. Then there exist two distributions π_0 and π_1 on $(\Omega, \mathscr{A})$ such that for every test Φ of Θ_0 against Θ_1 we have

$$R_\Phi^Q = C + A \int_\Omega \Phi\, d\pi_0 - B \int_\Omega \Phi\, d\pi_1$$

where C is a constant.

Proof The constants A and B are positive. Thus we can define the two distributions Q_0 and Q_1 by

$$\frac{dQ_0}{dQ}(\theta) = \begin{cases} \dfrac{1}{A}[W(\theta,1) - W(\theta,0)], & \text{if } \theta \in \Theta_0, \\ 0, & \text{if not;} \end{cases}$$

$$\frac{dQ_1}{dQ}(\theta) = \begin{cases} \dfrac{1}{B}[W(\theta,0) - W(\theta,1)], & \text{if } \theta \in \Theta_1, \\ 0, & \text{if not.} \end{cases}$$

On the other hand,

$$R_\Phi^Q = \iint_{\Omega \times (\Theta_0 \cup \Theta_1)} [W(\theta, 0) + [W(\theta, 1) - W(\theta, 0)]\Phi(\omega)] \, dP_\theta(\omega) \, dQ(\theta) \quad (2)$$

and we define π_0 and π_1 by

$$\forall A \in \mathscr{A}, \qquad \pi_0(A) = \int_{\Theta_0} P_\theta(A) \, dQ_0(\theta), \qquad \pi_1(A) = \int_{\Theta_1} P_\theta(A) \, dQ_1(\theta).$$

Using Theorem A.6.1 below and integrating (2) with respect to θ, the theorem is proved and the constant C is equal to

$$C = \int_{\Theta_0 \cup \Theta_1} W(\theta, 0) \, dQ(\theta). \qquad \blacksquare$$

Remark This theorem shows that if a loss function and a prior distribution are given, testing Θ_0 against Θ_1 is reduced to testing two simple hypotheses, i.e., involving a single distribution each.

To every quasi-ordering on tests we associated the notion of *admissibility* and the notion of *complete class* as introduced in Section 4.4. When we do not specify the quasi-ordering we shall mean the quasi-ordering in Definition 1.

Definition 2 We say that the test Φ of $\mathscr{P}_0$ against $\mathscr{P}_1$ is *quasi-admissible* if for every test Φ' of $\mathscr{P}_0$ against $\mathscr{P}_1$ such that

$$\beta_{\Phi'}(P) \leqslant \beta_\Phi(P), \qquad \forall P \in \mathscr{P}_0,$$

either we have

$$\beta_\Phi(P) = \beta_{\Phi'}(P), \qquad \forall P \in \mathscr{P}_1,$$

or there exists a distribution $P_1 \in \mathscr{P}$ such that

$$\beta_{\Phi'}(P_1) < \beta_\Phi(P_1).$$

We recall that a test is admissible if there does not exist another test which is better in the sense of Definition 1.

It is obvious that an admissible test of $\mathscr{P}_0$ against $\mathscr{P}_1$ is quasi-admissible but the converse is not necessarily true. The following properties are easily proven.

Theorem 3 *If Φ is an admissible test of $\mathscr{P}_0$ against $\mathscr{P}_1$, then $1 - \Phi$ is an admissible test of $\mathscr{P}_1$ against $\mathscr{P}_0$.*

Theorem 4 *Let $\mathscr{P}_0'$ be a nonempty subset of $\mathscr{P}_0$. If Φ is a quasi-admissible test of $\mathscr{P}_0'$ against $\mathscr{P}_1$, it is a quasi-admissible test of $\mathscr{P}_0$ against $\mathscr{P}_1$.*

Theorem 5 *If Φ is a quasi-admissible test of $\mathscr{P}_0$ against $\mathscr{P}_1$ and if $1 - \Phi$ is a quasi-admissible test of $\mathscr{P}_1$ against $\mathscr{P}_0$, then Φ is an admissible test of $\mathscr{P}_0$ against $\mathscr{P}_1$.*

3. Optimal Tests

In general, for a given hypothesis testing problem, there does not exist a test which is better than all others. If such a test did exist it would have to satisfy

$$\beta_\Phi(P) = 0, \qquad \forall P \in \mathscr{P}_0,$$

$$\beta_\Phi(P) = 1, \qquad \forall P \in \mathscr{P}_1.$$

But this condition implies that the set where $\Phi = 1$ (or $\Phi = 0$) is $\mathscr{P}_0$-negligible (or $\mathscr{P}_1$-negligible) and this condition is not fulfilled in general.

Definition 1 The test Φ of $\mathscr{P}_0$ against $\mathscr{P}_1$ is *uniformly most powerful* (u.m.p.), if for every test Φ' of $\mathscr{P}_0$ against $\mathscr{P}_1$ we have

$$\alpha_{\Phi'} \leqslant \alpha_\Phi \quad \Rightarrow \quad \beta_{\Phi'}(P) \leqslant \beta_\Phi(P), \qquad \forall P \in \mathscr{P}_1.$$

If the test Φ is u.m.p. and admissible we say that it is *strictly* u.m.p.

The following shows that a u.m.p. test is quasi-admissible:

$$\beta_{\Phi'}(P) \leqslant \beta_\Phi(P), \quad \forall P \in \mathscr{P}_0 \;\Rightarrow\; \alpha_{\Phi'} \leqslant \alpha_\Phi \;\Rightarrow\; \beta_{\Phi'}(P) \leqslant \beta_\Phi(P), \quad \forall P \in \mathscr{P}_1.$$

But a u.m.p. test is not necessarily admissible. A u.m.p. test is unbiased since it is better than the trivial test having the same level of significance.

Theorem 1 *Let $\mathscr{P}'_0$ be a subset of $\mathscr{P}_0$ and let Φ be a test of $\mathscr{P}_0$ against $\mathscr{P}_1$ having α_Φ as level of significance. If, as a test of $\mathscr{P}'_0$ against $\mathscr{P}_1$, Φ is u.m.p. and has a level of significance equal to α_Φ, then Φ is u.m.p. as a test of $\mathscr{P}_0$ against $\mathscr{P}_1$.*

Proof Let Φ' be a test of $\mathscr{P}_0$ against $\mathscr{P}_1$ such that

$$\alpha_{\Phi'} \leqslant \alpha_\Phi.$$

We have

$$\sup(\beta_{\Phi'}(P) \mid P \in \mathscr{P}'_0) \leqslant \sup(\beta_{\Phi'}(P) \mid P \in \mathscr{P}_0) = \alpha_{\Phi'} \leqslant \alpha_\Phi$$

and then

$$\beta_{\Phi'}(P) \leqslant \beta_\Phi(P), \qquad \forall P \in \mathscr{P}_1. \qquad \blacksquare$$

Theorem 2 *A test Φ of $\mathscr{P}_0$ against $\mathscr{P}_1$ is u.m.p. if and only if, for every distribution $P_1 \in \mathscr{P}_1$, Φ is a u.m.p. test of $\mathscr{P}_0$ against P_1.*

This theorem is a direct consequence of the definition.

Theorem 3 *If a free test Φ of $\mathscr{P}_0$ against the simple hypothesis P_1 is quasi-admissible* (*or admissible*) *then Φ is a u.m.p.* (*or strictly u.m.p.*) *test of $\mathscr{P}_0$ against $\mathscr{P}_1$.*

Proof The test Φ is free. If Φ' is another test we have

$$\alpha_{\Phi'} \leqslant \alpha_\Phi \quad \Rightarrow \quad \beta_{\Phi'}(P) \leqslant \beta_\Phi(P), \qquad \forall P \in \mathscr{P}_0.$$

If Φ is quasi-admissible then we have

$$\beta_{\Phi'}(P_1) \leqslant \beta_\Phi(P_1). \qquad \blacksquare$$

Similarly, if Φ is admissible, it is strictly u.m.p.

Theorem 4 *Let Φ be a test of $\mathscr{P}_0$ against $\mathscr{P}_1$ such that for every two distributions P_0 and P_1 belonging to $\mathscr{P}_0$ and $\mathscr{P}_1$ respectively, Φ is a quasi-admissible test of P_0 against P_1. Then Φ is a u.m.p. test of $\mathscr{P}_0$ against $\mathscr{P}_1$.*

Proof Using Theorem 2 we can restrict ourselves to the case of a simple hypothesis, $\mathscr{P}_1 = \{P_1\}$. For every distribution P_0 belonging to $\mathscr{P}_0$, Φ is a quasi-admissible test of P_0 against P_1. Thus if Φ' is a test of $\mathscr{P}_0$ against P_1 we have

$$\beta_{\Phi'}(P_0) \leqslant \beta_\Phi(P_0) \quad \Rightarrow \quad \beta_{\Phi'}(P_1) \leqslant \beta_\Phi(P_1). \tag{1}$$

On the other hand, if

$$\alpha_{\Phi'} \leqslant \alpha_\Phi,$$

there exists a distribution P_0 belonging to $\mathscr{P}_0$ such that

$$\beta_{\Phi'}(P_0) \leqslant \beta_\Phi(P_0).$$

Then, using (1), we see that Φ is u.m.p. $\blacksquare$

Definition 2 The test Φ of $\mathscr{P}_0$ against $\mathscr{P}_1$ is *uniformly most powerful unbiased* (u.m.p.u.) if for every unbiased test Φ' of $\mathscr{P}_0$ against $\mathscr{P}_1$,

$$\alpha_{\Phi'} \leqslant \alpha_\Phi \quad \Rightarrow \quad \beta_{\Phi'}(P) \leqslant \beta_\Phi(P), \qquad \forall P \in \mathscr{P}_1.$$

If Φ is u.m.p.u. and admissible we say that it is *strictly* u.m.p.u.

It is obvious that a u.m.p. test is u.m.p.u. Furthermore, the following argument shows that a u.m.p.u. test is unbiased and quasi-admissible. If there exists a test Φ' such that

$$\beta_{\Phi'}(P) \leqslant \beta_{\Phi}(P), \qquad \forall P \in \mathscr{P}_0.$$

$$\beta_{\Phi'}(P) \geqslant \beta_{\Phi}(P), \qquad \forall P \in \mathscr{P}_1,$$

and if one of these inequalities is strict for a distribution P_1 belonging to $\mathscr{P}_1$ this test Φ' would be unbiased since Φ is also unbiased. Then the previous inequalities would be inconsistent with the fact that Φ is a u.m.p.u. test.

Remark The test Φ of $\mathscr{P}_0$ against $\mathscr{P}_1$ is u.m.p.u. *if and only if*, for every distribution P_1 belonging to $\mathscr{P}_1$, $\beta_{\Phi}(P_1)$ is the maximum of $\beta_{\psi}(P_1)$ among all unbiased tests ψ of $\mathscr{P}_0$ against $\mathscr{P}_1$ (and not P_1 !!).

Definition 3 The test Φ of $\mathscr{P}_0$ against $\mathscr{P}_1$ is *maximin*, if for every test Φ',

$$\alpha_{\Phi'} \leqslant \alpha_{\Phi} \quad \Rightarrow \quad \inf(\beta_{\Phi'}(P) \mid P \in \mathscr{P}_1) \leqslant \inf(\beta_{\Phi}(P) \mid P \in \mathscr{P}_1).$$

It is easy to see that a test which is u.m.p. or u.m.p.u. is a maximin one. On the other hand, if $\mathscr{P}_1$ is a simple hypothesis P_1, the three properties, u.m.p., u.m.p.u., and maximin are equivalent. Finally, as in the case of the other properties, a maximin test is unbiased and quasi-admissible.

Theorem 5 *Let* $(\Omega, \mathscr{A}, \{P_\theta, \theta \in (\Theta, \mathscr{T})\})$ *be a dominated statistical space and let* Θ_0 *and* Θ_1 *be two disjoint and measurable hypotheses. For every real number* α *belonging to* $]0,1[$ *there exists a maximin test* Φ *of* Θ_0 *against* Θ_1 *having level of significance* α. *This result remains true if we restrict ourselves to free tests.*

Proof Let $\mathscr{R}$ be the space associated with this problem. Theorem 12.3.5 will show that $\mathscr{R}$ is a compact set of $\mathscr{L}_\infty(\Theta', \mathscr{T} \cap \Theta')$, $(\Theta' = \Theta_0 \cup \Theta_1)$, with respect to the topology $\sigma\{\mathscr{L}_\infty, \mathscr{M}\}$. Let $\mathscr{R}_\alpha$ be the subset of $\mathscr{R}$ defined by

$$f \in \mathscr{R}_\alpha \quad \Leftrightarrow \quad f \in \mathscr{R}, f(\theta) \leqslant \alpha, \qquad \forall \theta \in \Theta_\theta.$$

$\mathscr{R}_\alpha$ is nonempty since the trivial level of significance α-test belongs to $\mathscr{R}$. Furthermore, $\mathscr{R}_\alpha$ is compact since it is the intersection of $\mathscr{R}$ with a family of closed half-spaces. The mapping f from $\mathscr{L}_\infty$ in $\mathbb{R}$ defined by

$$f \in \mathscr{L}_\infty, \qquad f \to \inf(f(\theta) \mid \theta \in \Theta_1),$$

is an upper semicontinuous one. Then there exists an element f^* belonging to $\mathscr{L}_\infty$ such that

$$\inf(f^*(\theta) \mid \theta \in \Theta_1) = \sup[\inf(f(\theta) \mid \theta \in \Theta_1) \mid f \in \mathscr{R}_\alpha].$$

The element f^* is the image of a maximin test Φ of Θ_0 against Θ_1. The proof is completed by showing that the level of significance of Φ is α, i.e.,

$$\sup(f^*(\theta)\,|\,\theta \in \Theta_0) = \alpha.$$

This will be shown by a *reductio ad absurdum* argument. If the previous condition is not satisfied there exists a number ε belonging to $]0, 1[$ such that

$$g = f^*(1 - \varepsilon) + \varepsilon$$

belongs to $\mathscr{R}_\alpha$ since it is a convex combination of Φ and of 1_Ω. Then we would have

$$\inf(g(\theta)\,|\,\theta \in \Theta_1) = (1 - \varepsilon)\inf(f^*(\theta)\,|\,\theta \in \Theta_1) + \varepsilon.$$

Using the definition of f^*, we have

$$\inf(f^*(\theta)\,|\,\theta \in \Theta_1) = 1.$$

Thus g has the same property and we can choose ε such that

$$\sup(g(\theta)\,|\,\theta \in \Theta_0) = \alpha.$$

If we restrict ourselves to free tests, the argument is identical, if we replace $\mathscr{R}_\alpha$ by $\mathscr{R}'_\alpha$, defined by

$$f \in \mathscr{R}'_\alpha \quad \Leftrightarrow \quad f(\theta) = \alpha, \qquad \forall\theta \in \Theta_0. \qquad \blacksquare$$

Remark If Θ_1 contains a single point θ_1, this theorem establishes the existence of a u.m.p. test having a given level of significance.

Theorem 6 (*Least Favorable Distributions*) *Let $(\Omega, \mathscr{A}; \{P_\theta, \theta \in (\Theta, \mathscr{T})\})$ be a statistical space, Θ_0 a measurable subset of Θ, θ_1 an element of $\Theta - \Theta_0$ and Q a distribution on $(\Theta_0, \Theta_0 \cap \mathscr{T})$. If Φ is a quasi-admissible test of*

$$P_0^* = \int_{\Theta_0} P_\theta \, dQ(\theta)$$

against P_{θ_1}, then it is a quasi-admissible test of Θ_0 against θ_1. Furthermore if

$$Q(\{\theta : \beta_\Phi(\theta) = \alpha_\Phi, \theta \in \Theta_0\}) = 1, \tag{2}$$

Φ is a u.m.p. test of Θ_0 against θ_1.

Proof Let Φ' be a test of Θ_0 against θ_1 such that

$$\beta_{\Phi'}(\theta) \leqslant \beta_\Phi(\theta), \qquad \forall\theta \in \Theta_0.$$

We have

$$\int_{\Theta_0} \beta_{\Phi'}(\theta)\, dQ(\theta) \leqslant \int_{\Theta_0} \beta_\Phi(\theta)\, dQ(\theta).$$

Using Theorems 1.4.2 and A.6.1 we have

$$\int_{\Omega} \Phi' \, dP_0^* \leqslant \int_{\Omega} \Phi \, dP_0^*,$$

then

$$\beta_{\Phi'}(\theta_1) \leqslant \beta_{\Phi}(\theta_1).$$

Thus, Φ is a quasi-admissible test of Θ_0 against θ_1.

Now, let Θ_0' be the subset of Θ_0 defined by

$$\beta_{\Phi}(\theta) = \alpha_{\Phi}, \qquad \theta \in \Theta_0.$$

Using (2) we have

$$P_0^* = \int_{\Theta_0'} P_\theta \, dQ(\theta).$$

The previous result shows that Φ is a quasi-admissible and free test of Θ_0' against θ_1 and thus is u.m.p. Finally, Φ has the same level of significance as a test of Θ_0 against θ_1 or as a test of Θ_0' against θ_1. ∎

Remark The distributions P_θ, where θ belongs to Θ_0' are called *least favorable distributions.*

4. The Fundamental Neyman–Pearson Lemma

Theorem 1 *Let μ be a positive measure on the measurable space $(\Omega, \mathscr{A})$ and let $f_1, \ldots, f_k$ be k real and μ-integrable functions on $(\Omega, \mathscr{A})$. To every measurable function Φ which is defined on $(\Omega, \mathscr{A})$ taking values in $[0, 1]$, we associate the point belonging to $\mathbb{R}^k$ having components*

$$\beta_\Phi^1 = \int \Phi f_1 \, d\mu, \ldots, \beta_\Phi^k = \int \Phi f_k \, d\mu.$$

The locus of the point S_Φ is a convex bounded and closed subset D of $\mathbb{R}^k$. If S_Φ belongs to the boundary of D there exist constants $a_1, \ldots, a_k$ such that

$$\begin{cases} \sum_{j=1}^{k} f_j(\omega) a_j > 0 \ \Rightarrow \ \Phi = 1, & \mu\text{-a.e.} \\ \sum_{j=1}^{k} f_j(\omega) a_j < 0 \ \Rightarrow \ \Phi = 0, & \mu\text{-a.e.} \end{cases}$$

If S_Φ is an extremal point of D and if the given functions $f_1 \cdots f_k$ are positive there exists a set A belonging to $\mathscr{A}$ such that

$$S_\Phi = S_{1_A}.$$

Proof It is obvious that D is convex. For every function f_j we write f_j^+ and f_j^- for its positive and negative components, i.e.,

$$f_j = f_j^+ - f_j^-,$$

and let P_j^+ and P_j^- be the distributions

$$dP_j^+ = f_j^+ \, d\mu \Big/ \int_\Omega f_j^+ \, d\mu, \qquad dP_j^- = f_j^- \, d\mu \Big/ \int_\Omega f_j^- \, d\mu, \qquad j = 1, \ldots, k.$$

Theorem 12.3.5 will show that D is compact with respect to a weak topology which, in this case, is the usual topology on $\mathbb{R}^k$.

Let S_Φ be a point belonging to the boundary of D; there exists at least one hyperplane of $\mathbb{R}^k$ containing S_Φ such that D is entirely on one side of this hyperplane. That is, there exist some constants $a_1, \ldots, a_k$ such that

$$\sum_1^k a_j(\beta_\Phi^j - x_j) \geqslant 0, \qquad \forall (x_1, \ldots, x_k) \in D.$$

This is equivalent to

$$\sum_{j=1}^k a_j \beta_\Phi^j \geqslant \sum_{j=1}^k a_j \beta_{\Phi'}^j, \tag{1}$$

for every measurable function Φ' taking values on $[0, 1]$.

Let f^* be the function $\sum_1^k f_j a_j$ and let Φ^* be the indicator function of the set $\{f^* > 0\}$. We have

$$(\Phi^* - \Phi')f^* \geqslant 0. \tag{2}$$

Then integrating with respect to μ, we obtain

$$\int_\Omega \Phi^* f^* \, d\mu \geqslant \int_\Omega \Phi' f^* \, d\mu, \qquad \forall \Phi'. \tag{3}$$

Using (1) we deduce from (3) that

$$\int_\Omega \Phi^* f^* \, d\mu = \int_\Omega \Phi f^* \, d\mu = \sup_{\Phi'} \left(\int_\Omega \Phi' f^* \, d\mu \right)$$

and, using (2) we obtain

$$(\Phi^* - \Phi) f^* = 0, \qquad \mu\text{-a.e.},$$

i.e.,

$$\begin{cases} f^* > 0 \quad \Rightarrow \quad \Phi^* = \Phi = 1, & \mu\text{-a.e.,} \\ f^* < 0 \quad \Rightarrow \quad \Phi^* = \Phi = 0, & \mu\text{-a.e.} \end{cases}$$

Finally, let S_Φ be an extremal point of D and let μ^* be the positive measure defined by

$$d\mu^* = (f_1 + \cdots + f_k)\, d\mu.$$

If we have

$$\mu^*(\Phi(1 - \Phi) > 0) = 0, \tag{4}$$

then using the set A, where $\Phi = 1$, completes the proof. If condition (4) is not satisfied there exists an integer n such that the set

$$A_n = \left\{\omega : \Phi(I - \Phi) > \frac{1}{n}\right\}$$

does not have zero μ^*-measure. The two functions $\Phi + n^{-1}1_{A_n}$ and $\Phi - n^{-1}1_{A_n}$ have values on [0, 1] and their images are $S_\Phi + n^{-1}S_{1_{A_n}}$ and $S_\Phi - n^{-1}S_{1_{A_n}}$ respectively. Since $S_{1_{A_n}}$ is not zero, S_Φ cannot be an extremal point of D and the proof is completed by *reductio ad absurdum*. ∎

Theorem 2 (Neyman–Pearson Lemma) *Let μ be a positive measure on a measurable space $(\Omega, \mathscr{A})$ and let $f_0, f_1, \ldots, f_k$ be $k + 1$ real and μ-integrable functions defined on $(\Omega, \mathscr{A})$. If a function Φ takes values in [0, 1] and is such that*

$$f_0(\omega) > \sum_1^k f_j(\omega)a_j \quad \Rightarrow \quad \Phi = 1, \qquad \mu\text{-a.e.}$$

$$f_0(\omega) < \sum_1^k f_j(\omega)a_j \quad \Rightarrow \quad \Phi = 0, \qquad \mu\text{-a.e.,}$$

where the constants a_j are nonnegative, then for every function Φ', taking values on [0, 1], we have

$$\int_\Omega \Phi' f_j\, d\mu \leqslant \int_\Omega \Phi f_j\, d\mu, \qquad \forall j = 1, \ldots, k \quad \Rightarrow \quad \int_\Omega \Phi' f_0\, d\mu \leqslant \int_\Omega \Phi f_0\, d\mu. \tag{5}$$

This result remains true if the functions $f_1, \ldots, f_k$ are positive and if the functions f_0, $\Phi f_j (j = 1, \ldots, k)$ are μ-integrable.

Proof It is easily seen that the following function is positive

$$H(\omega) = [\Phi(\omega) - \Phi'(\omega)]\left(f_0(\omega) - \sum_1^k f_j(\omega)a_j\right) \geqslant 0.$$

If the functions $f_0, \ldots, f_k$ are integrable the function H is also integrable. If the assumptions of the second part of the theorem are satisfied we have

$$f_0(\omega) > \sum_1^k f_j(\omega) a_j \quad \Rightarrow \quad H(\omega) \leqslant f_0(\omega)$$

$$f_0(\omega) < \sum_1^k f_j(\omega) a_j \quad \Rightarrow \quad H(\omega) \leqslant \sum_1^k \Phi' f_j a_j - \Phi' f_0.$$

But the functions $\Phi' f_j$ are positive. Thus (5) shows that they are integrable. We can integrate H with respect to μ and obtain

$$\int_\Omega \Phi f_0 \, d\mu - \int_\Omega \Phi' f_0 \, d\mu \geqslant \sum_1^k a_j \left(\int_\Omega \Phi f_j \, d\mu - \int_\Omega \Phi' f_j \, d\mu \right) \geqslant 0,$$

and the proof follows easily from this inequality. ∎

Remark If in (5) we change the direction of one inequality, we must change the sign of the corresponding constant a_j. On the other hand, if one of these inequalities becomes an equality, the corresponding constant a_j can be positive or negative. For a converse of Theorem 2, see Section 12.5.

5. Determining Optimal Tests

When a prior distribution is given, the following theorem shows that we can find the Bayes solution of a hypothesis testing problem.

Theorem 1 *Using notation and assumptions of Theorem 2.2, there exists a deterministic test Φ such that for every test Φ' of Θ_0 against Θ_1 we have*

$$R_{\Phi'}^Q \geqslant R_\Phi^Q.$$

Furthermore this test is **admissible** *as a test of π_0 against π_1, and is called a* **Bayes-optimal** *test.*

Proof Let Δ be the locus in $[0, 1]^2$ of the points having coordinates

$$\left(\int_\Omega \Phi' \, d\pi_0, \ \int_\Omega \Phi' \, d\pi_1 \right),$$

where Φ' runs through all possible tests. Using Theorem 4.1, one can show that Δ is a closed convex subset. Then, the linear function R_Φ^Δ, which is defined on Δ is maximum at an extremal point of Δ. Let Φ be a test which corresponds to this extremal point. We have seen that the constants A and B which define R_Φ^ϱ are positive. Then Φ is on the arc (A) of the boundary of Δ (see Fig. 1), and we have

$$\int_\Omega \Phi \, d\pi_0 \leqslant \int_\Omega \Phi \, d\pi_1 .$$

Thus Φ is admissible. Finally, using the last part of Theorem 4.1 we can choose for Φ a deterministic test. ∎

Theorem 2 *Let P_0 and P_1 be two distributions on $(\Omega, \mathscr{A})$; let p_0 and p_1 be their respective densities with respect to the same positive measure μ. For every nonnegative constant λ, any test Φ such that*

$$p_1(\omega) > \lambda p_0(\omega) \quad \Rightarrow \quad \Phi = 1,$$
$$p_1(\omega) < \lambda p_0(\omega) \quad \Rightarrow \quad \Phi = 0,$$

is a u.m.p. test of P_0 against P_1. Also, for every $\alpha \in [0, 1]$ there exists one such test Φ having the level of significance α, and if $\beta_\Phi(P_1) < 1$ then the test Φ is strictly u.m.p.

The region Δ in Fig. 1 illustrates these two theorems. Note that the locus of the images of all admissible tests is the subset of the boundary of Δ which is above the diagonal of the square.

We can even compute completely a u.m.p. test having level of significance α. Let $G(\lambda)$ be the function defined by

$$\lambda \geqslant 0, \qquad G(\lambda) = P_0(\{p_1 > \lambda p_0\}).$$

We see easily that this function is continuous to the right and nonincreasing.

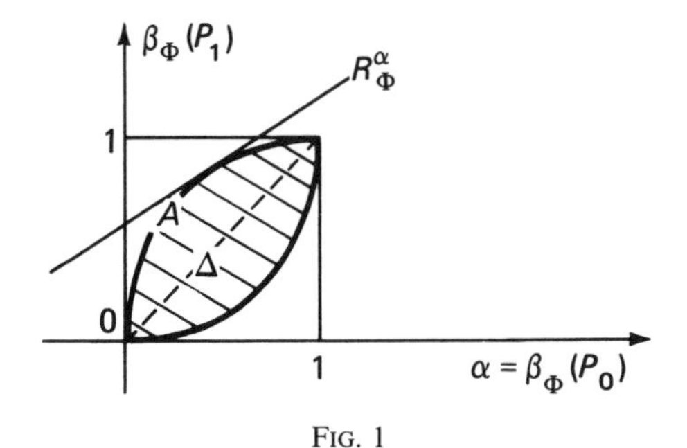

FIG. 1

Let α be given in $[0, 1]$. Then only one of the following cases can occur:

(a) $G(\lambda - 0) = G(\lambda) = \alpha$. Then every test such that

$$\Phi = 1, \quad \text{if} \quad p_1 > \lambda p_0,$$
$$\Phi = 0, \quad \text{if} \quad p_1 < \lambda p_0,$$

is a u.m.p. deterministic test having level of significance α.

(b) $G(\lambda) \leqslant \alpha < G(\lambda - 0)$. Then the test

$$\Phi = \begin{cases} 1 & \text{if} \quad p_1 > \lambda p_0 \\ 0 & \text{if} \quad p_1 < \lambda p_0 \\ \dfrac{\alpha - G(\lambda)}{G(\lambda - 0) - G(\lambda)} & \text{if} \quad p_1 = \lambda p_0 \end{cases}$$

is u.m.p. and has level of significance α.

Theorem 3 *Let $P_0, P_1, \ldots, P_N$ be $N + 1$ distributions on $(\Omega, \mathscr{A})$ and let $p_0, \ldots, p_N$ be their densities with respect to a given positive measure. For every set of nonnegative (positive) constants $\lambda_0, \ldots, \lambda_{N-1}$ every test Φ such that*

$$p_N(\omega) > \sum_{j=0}^{N-1} \lambda_j p_j(\omega) \quad \Rightarrow \quad \Phi(\omega) = 1,$$

$$p_N(\omega) < \sum_{j=0}^{N-1} \lambda_j p_j(\omega) \quad \Rightarrow \quad \Phi(\omega) = 0,$$

is a quasi-admissible (admissible) test of $\mathscr{P}_0 = \{P_0, \ldots, P_{N-1}\}$ against P_N.

Proof When the constants $\lambda_0, \ldots, \lambda_{N-1}$ are nonnegative, this theorem is a direct application of the Neyman–Pearson Lemma. In the case of strictly positive constants λ_j the following argument shows that the test Φ is admissible. We have seen that

$$\beta_\Phi(P_N) - \beta_{\Phi'}(P_N) \geqslant \sum_{j=0}^{N-1} \lambda_j[\beta_\Phi(P_j) - \beta_{\Phi'}(P_j)].$$

Then if one of the inequalities

$$\beta_\Phi(P_j) \geqslant \beta_{\Phi'}(P_j), \qquad j = 0, \ldots, N - 1,$$

is strictly realized, we have

$$\beta_\Phi(P_N) > \beta_{\Phi'}(P_N),$$

which shows that Φ is admissible. ∎

Remark Theorem 12.4.2 will show that if the distributions $P_0, \ldots, P_N$ are nonatomic, then to every test Φ there corresponds an equivalent deterministic test.

To seek a u.m.p. test of Θ_0 against Θ_1, when the previous theorems cannot be applied, use the following procedure:

(a) Arbitrarily choose a value θ_0 belonging to Θ_0 and a value θ_1 belonging to Θ_1. Using Theorem 2, set up the u.m.p. tests of θ_0 against θ_1.

(b) If Theorem 3.4 can be applied, the problem is solved. If not, keep the value θ_1 but look for a subset Θ_0' of Θ_0 (least favorable hypotheses) such that by using Theorems 2 or 3 a quasi-admissible free test of Θ_0' against θ_1 can be found. If it can be verified that this test Φ has the same level of significance (considered as a test of Θ_0 against θ_1), then Theorem 3.6 implies that Φ is u.m.p.

(c) If for every θ_1 in Θ_1, a u.m.p. test of Θ_0 against θ_1 can be found, then by using Theorem 3.2 determine the existence or nonexistence of a u.m.p. test of Θ_0 against Θ_1.

Remark When the hypothesis θ_1 is simple, Theorem 3.5 shows that there exists a u.m.p. test. If every trivial test is u.m.p. (for example, when θ_1 belongs to $\overline{\Theta}_0$ in Theorem 1.3) we look for a strictly u.m.p. test.

If no u.m.p. test exists, we can look for a u.m.p.u. test. To do this, consider the following inequalities (unbiasedness)

$$\begin{aligned} \beta_\Phi(\theta) &\leqslant \alpha, \qquad \forall \theta \in \Theta_0, \\ \beta_\Phi(\theta) &\geqslant \alpha, \qquad \forall \theta \in \Theta_1. \end{aligned} \tag{1}$$

For every θ_1 belonging to Θ_1 we seek the maximum of $\beta_\Phi(\theta_1)$ when these inequalities are satisfied. In the case of a dominated statistical space this maximum exists, and a test Φ is u.m.p.u. if and only if for every θ_1 belonging to Θ_1 this maximum is reached for the test Φ.

In general, inequalities (1) are awkward to handle. Thus one method consists of weakening these inequalities. For example, if Theorem 1.3 can be applied, we replace (1) by

$$\beta_\Phi(\theta) = \alpha, \qquad \forall \theta \in \overline{\Theta}_0 \cap \overline{\Theta}_1.$$

In the case where (1) implies that for a value θ_0 we have

$$d\beta_\Phi(\theta_0)/d\theta = 0,$$

we replace (1) by this equality. Both of these methods will be applied in the exercises and in Chapter 11. Note that, in many cases, the Neyman–Pearson Lemma, can be applied with the equality constraints since they are linear functionals of Φ.

In some cases, the following theorem (method of conditional tests) can eliminate the nuisance parameters. This is an application of the concept introduced in Section 4.6 where we replace a nuisance parameter by a conditioning variable.

Theorem 4 *Let* $[\mathscr{X} \times \mathscr{Y}, \mathscr{A} \otimes \mathscr{B}, \{P_\theta, \theta = (\lambda, \mu) \in \Lambda \times M\}]$ *be a statistical space such that, for every* $\theta \in \Theta$, *the distribution on* $\mathscr{A}$ *given* Y *exists and does not depend on* μ. *Let* P_λ^y *be this conditional distribution. Consider a test* $\Phi^*(x, y)$, *such that for every* $y \in \mathscr{Y}$, *the test* Φ_y^* *of* Λ_0 *against* Λ_1 *defined on the statistical space,*

$$[\mathscr{X}, \mathscr{A}; P_\lambda^y, \lambda \in \Lambda] \qquad \textit{by} \qquad \Phi_y^* : x \to \Phi^*(x, y)$$

is better than every free test of Λ_0 *against* Λ_1 *having the same level of significance. Then the test* Φ^* *on the given statistical space is better than every test of* $\Lambda_0 \times M$ *against* $\Lambda_1 \times M$, *which has the same level of significance and is free given* Y.

Proof In this theorem Y is the projection of $\mathscr{X} \times \mathscr{Y}$ on $\mathscr{Y}$; that is $(x, y) \to y$. For every test Φ we have

$$E_{P_\theta}(\Phi \mid Y) = \int_{\mathscr{X}} \Phi(x, y)\, dP_\lambda^y(x).$$

Then if Φ is free given Y on $\Lambda_0 \times M$ and has the level of significance α, we have

$$\forall y \in \mathscr{Y}, \quad \forall \lambda \in \Lambda_0, \qquad \int_{\mathscr{X}} \Phi(x, y)\, dP_\lambda^y(x) = \alpha.$$

That is, the test $x \to \Phi(x, y)$ which is defined on the statistical space

$$(\mathscr{X}, \mathscr{A}; P_\lambda^y, \lambda \in \Lambda_0)$$

is free and has level of significance α. Then

$$\forall y \in \mathscr{Y}, \quad \forall \lambda \in \Lambda_1, \qquad \int_{\mathscr{X}} \Phi(x, y)\, dP_\lambda^y(x) \leqslant \int_{\mathscr{X}} \Phi^*(x, y)\, dP_\lambda^y(x).$$

Integrating with respect to y we have

$$\beta_\Phi(\theta) \leqslant \beta_{\Phi^*}(\theta), \qquad \forall \theta \in \Lambda_1 \times M.$$

On the other hand, Φ is free and has level of significance α. Thus

$$\forall \theta \in \Lambda_0 \times M, \qquad \beta_\Phi(\theta) = \alpha \geqslant \beta_{\Phi^*}(\theta),$$

since Φ^* has level of significance α. ∎

Note that when every unbiased test is free given Y (see, e.g., Section 11.4), this theorem can be used to establish the u.m.p.u. property.

6. Nonoptimal Methods

In most cases an optimal test does not exist. This is often true in the parametric case (dim Θ finite) and is always true in the usual nonparametric case (dim Θ infinite). We now indicate some of the methods used when one cannot find an optimal test.

A. Restrict the Class of Tests Which Are Taken into Consideration

Let $(\Omega, \mathscr{A}, \mathscr{P})$ be the given statistical space. We can restrict consideration to $\mathscr{B}$-measurable tests where $\mathscr{B}$ is a given subfield of $\mathscr{A}$. This method consists in replacing the given statistical space by $(\Omega, \mathscr{B}, \mathscr{P})$. As an example of the application of this method consider the study of homogeneous tests for the Behrens–Fisher problem (Section 10.5). We can also consider tests whose power function is $\mathscr{C}$-measurable where $\mathscr{C}$ is a given subfield of the σ-field given on the parameter space. This is the method of invariance as developed by Linnik [41, 42]. However it is possible that the power function of every test is not $\mathscr{C}$-measurable. Thus the results obtained by this method are rather negative. Finally, we can consider tests which are invariant with respect to a group of transformations on Ω for which the hypotheses $\mathscr{P}_0$ and $\mathscr{P}_1$ remain the same [44; p. 215].

B. Likelihood Ratio Test

Let $\mathscr{L}(\omega, \theta)$ be the likelihood function of the given statistical space. The test Φ of Θ_0 against Θ_1 is called a *likelihood ratio test* if there exists a constant λ such that

$$\frac{\sup\{\mathscr{L}(\omega, \theta) \mid \theta \in \Theta_0\}}{\sup\{\mathscr{L}(\omega, \theta) \mid \theta \in \Theta_1\}} > \lambda \quad \Rightarrow \quad \Phi = 0,$$

$$\frac{\sup\{\mathscr{L}(\omega, \theta) \mid \theta \in \Theta_0\}}{\sup\{\mathscr{L}(\omega, \theta) \mid \theta \in \Theta_1\}} < \lambda \quad \Rightarrow \quad \Phi = 1.$$

Lehman has stated [44, p. 15], "Although the maximum likelihood principle is not based on any clearly defined optimum considerations, it has been very successful in leading to satisfactory procedures in many specific problems. For wide classes of problems, maximum likelihood procedures have also been shown to possess various asymptotic optimum properties as the sample size tends to infinity."

In some cases, we shall study tests which are convex combinations of tests, where each of them has an optimal property. For example, suppose that

$$\mathscr{P}_1 = \bigcup_{i \in I} \mathscr{P}_1^i,$$

where I is finite. If for every i belonging to I there exists an optimal test Φ_i of $\mathscr{P}_0$ against $\mathscr{P}_i$, then the following test will be of interest

$$\Phi = \sum_{i \in I} \lambda_i \Phi_i \qquad \left(\lambda_i > 0, \ \sum_{i \in I} \lambda_i = 1\right).$$

C. The General Nonparametric Case

In this case, the methods above rarely yield results. Then one would use free statistics (or asymptotically free statistics) whose distributions are known. The classical tests are based on well-known theorems of probability theory which provide the basis for computing their levels of significance. For example, the chi-square test is based on the theorem of K. Pearson (Theorem 7.3.2) and the tests of Kolmogorov–Smirnov [17, 50], Smirnov [23], and Geary (Exercise 12.9) are also based on corresponding well-known theorems.

D. Asymptotic Study

Finally, in the case of a sample $(\Omega, \mathscr{A}, \mathscr{P})^n$, an asymptotic study is often useful. Let $\mathscr{P}_0$ and $\mathscr{P}_1$ be two hypotheses; we say that the sequence of tests Φ_n is a *consistent* one if

$$\alpha_{\Phi_n} \to 0, \qquad \beta_{\Phi_n}(P) \to 1, \qquad \forall P \in \mathscr{P}_1,$$

when n tends to infinity. The chi-square test and the Kolmogorov–Smirnov test are consistent.

Exercises

1. Let P_0 and P_1 be two given distributions on the finite set E. Suppose that we have one observation on a random variable taking values on E whose distribution is either P_0 or P_1.

(a) Write the statistical space corresponding to this statistical experiment.

(b) Using the Neyman–Pearson Lemma, find the admissible tests of P_0 against P_1.

(c) Show that the space associated with (b), is a closed and convex polygon.

(d) Find u.m.p. tests of P_0 against $\mathscr{P}_1$, where the family $\mathscr{P}_1$ consists of the distributions.

$$P_\lambda = \lambda P_0 + (1 - \lambda)P_1, \qquad \lambda \in [0, 1[.$$

2. Construct some examples of testing statistical hypotheses in which u.m.p. tests are not admissible.

3. Consider the statistical space $[\mathbb{R}, \mathscr{B}_{\mathbb{R}}; N(m, \sigma^2), m \in \mathbb{R}]$, and the two tests

$$\Phi_1 = 1_{\complement[a,b]}, \qquad \Phi_2 = 1_{\complement[c,d]}$$

where the four real numbers a, b, c, d satisfy $a < b < c < d$. Compute the power function of each test Φ_1 and Φ_2 of the hypothesis $\{m = m_0\}$ against $\{m \neq m_0\}$ and show that these two power functions have at least one common point.

4. How many independent observations on an event E, having the unknown probability p must be performed in order to be able to test the hypothesis $\{p = p_0\}$ against the hypothesis $\{p = p_1\}$ with given risks? Using the normal approximation to the binomial distribution find a rough estimate of this number.

5. Consider the statistical space $[\mathbb{R}^+, \mathscr{B}_{\mathbb{R}^+}; e^{-x/\theta}/\theta, \theta \in \mathbb{R}^+]^n$.

(a) Let θ_0 be a given value of θ. Find a test of $\{\theta = \theta_0\}$ against $\{\theta > \theta_0\}$. Compute its power function and level of significance.

(b) Solve the same problem if, instead of n observations we know only that the number of observations is greater than a given real number a.

(c) Consider the statistical space

$$[\mathbb{R}^+, \mathscr{B}_{\mathbb{R}^+}; e^{-x/\theta}/\theta, \theta \in \mathbb{R}^+]^n \otimes [\mathbb{R}^+, \mathscr{B}_{\mathbb{R}^+}; e^{-y/\theta'}/\theta', \theta' \in \mathbb{R}^+]^{n'}$$

and describe the statistical experiment which corresponds to this statistical space. Suggest a test of $\{\theta = \theta'\}$ against $\{\theta > \theta'\}$ and compute its power function.

6. Consider a sample of size n from the uniform distribution on $[0, \theta]$ where θ is an unknown parameter. Let a be a real number greater than 1.

(a) Find all deterministic admissible tests of $\{\theta = 1\}$ against $\{\theta = a\}$.

(b) Find all admissible tests of $\{\theta = 1\}$ against $\{\theta = a\}$ and express these tests by means of those found in (a).

(c) Describe the space associated with this problem of hypothesis testing.

7. Consider the statistical space

$$[\mathbb{R}^k, \mathscr{B}_{\mathbb{R}^k}; N(m, \Lambda), m \in \mathbb{R}^k]^n,$$

where Λ is known, and find the admissible tests of $\{m = m_0\}$ against $\{m = m_1\}$ where m_0 and m_1 are given. Find a hypothesis $\mathscr{P}_1$ such that the tests which have been previously found are u.m.p. tests of $\{m = m_0\}$ against $\mathscr{P}_1$.

8. Consider the sample

$$[\mathbb{R}, \mathscr{B}_{\mathbb{R}}; N(0, \theta^{-1/2}), \theta \in \mathbb{R}^+]^n.$$

(a) Find the u.m.p. test of $\{\theta = 1\}$ against $\{\theta > 1\}$ which has a given level of significance and compute its power function.

(b) Suppose that the unknown parameter θ takes only the values 1 and 2 with given prior probabilities. Assume that we reject the hypothesis $\{\theta = 1\}$ when its posterior probability is less than $\frac{1}{2}$. Find the corresponding test.

(c) Consider the loss function defined by

$$W(1, 2) = 1, \qquad W(2, 1) = 5.$$

Assume that the prior probability of $\{\theta = 1\}$ is $p = 0.8$. Compute the optimal test. What can you say about it?

9. Consider the following sample

$$\left[\mathbb{R}^+, \mathscr{B}_{\mathbb{R}^+}; \frac{\lambda^a}{\Gamma(a)} e^{-\lambda x} x^{a-1}, a \in \mathbb{R}^+\right]^n,$$

where λ is known. Let $\{a = a_0\}$ and $\{a = a_1\}$ be two simple hypotheses ($a_0 < a_1$) having the prior probabilities p_0 and p_1 ($p_0 + p_1 = 1$) respectively. We choose the hypothesis having the greatest posterior probability; what test of $\{a = a_0\}$ against $\{a = a_1\}$ do we obtain? Comment.

10. Consider a sample of size n on the normal distribution $N(m, 1)$. Find an optimal test of the hypothesis $\{m \neq 0\}$ against $\{m = 0\}$. Comment.

11. Consider a sample of size n from the Poisson distribution with unknown parameter. Find the optimal tests of the hypothesis $\{\lambda < \lambda_0\}$ against the hypothesis $\{\lambda > \lambda_0\}$. Show that the optimality property remains if one of the inequalities is not strict.

12. Consider the sample, $[\mathbb{R}, \mathscr{B}_{\mathbb{R}}; N(m, 1), m \in \mathbb{R}]^n$. Compute the integer n such that one can distinguish between the two hypotheses $\{m < -\varepsilon\}$, $\{m > \varepsilon\}$ (for given $\varepsilon > 0$) with a probability of error less than .001.

13. Consider the sample, $[\mathbb{R}, \mathscr{B}_{\mathbb{R}}; N(m, \sigma^2), m \in \mathbb{R}, \sigma \in \mathbb{R}^+]^n$. Find the optimal test of the simple hypothesis $\{m = m_0, \sigma = \sigma_0\}$ against the simple hypothesis $\{m = m_1, \sigma = \sigma_1\}$. Compute the level of significance and the power of this test. [Use the cumulative distribution function of the non-central χ^2-distribution.]

Suggest a test of $\{m = m_0, \sigma = \sigma_0\}$ against $\{(m, \sigma) \neq (m_0, \sigma_0)\}$. [Estimate m and σ by the maximum likelihood method.]

14. Consider the sample $[\mathbb{R}, \mathscr{B}_{\mathbb{R}}; N(m, 1), m \in \mathbb{R}]^n$ to be used to test the hypothesis $\{m = 0\}$ against $\{|m| = 1\}$. Let $\beta_\Phi(m)$ be the power function of the test Φ.

(a) Using the Neyman–Pearson Lemma in a suitable way, find the test having the level of significance α which maximizes the quantity

$$\tfrac{1}{2}[\beta_\Phi(1) + \beta_\Phi(-1)].$$

Compute the value of this maximum in terms of α.

(b) Show that the test which has been obtained in (a) is the α level of significance test which is most powerful among tests for which

$$\beta_\Phi(1) = \beta_\Phi(-1).$$

(c) Using the Neyman–Pearson Lemma in a suitable way, find the unbiased α level of significance test for which $\beta_\Phi(1)$ is maximum.

(d) Using the previous results and Exercise 3, show that no u.m.p.u. test exists.

(e) Find a u.m.p. test of the hypothesis $\{|m| = 1\}$ against $\{m = 0\}$. Comment.

15. Let P_θ be the distribution on $[-\frac{1}{2}, +\frac{1}{2}]$ which is defined by the density

$$\begin{cases} 1 - \theta & \text{if} \quad -\frac{1}{2} \leqslant x \leqslant 0, \\ 1 + \theta & \text{if} \quad 0 < x \leqslant \frac{1}{2}, \end{cases}$$

and consider a random sample of size n from the distribution P_θ.

(a) Find the optimal test of the hypothesis $\{\theta = 0\}$ against $\{0 < \theta \leqslant 1\}$ and suggest an unbiased test of $\{\theta = 0\}$ against $\{\theta \neq 0\}$.

(b) We have 10,000 independent observations on a random variable which is uniformly distributed on $[-\frac{1}{2}, +\frac{1}{2}]$. Suppose that some of the signs of the observations have been lost. For each observation this occurrence has a constant and unknown probability. Find a test of the hypothesis that this probability is zero against the alternative that it is not zero.

(c) Now suppose that, with a constant and unknown probability, the sign of each observation is damaged. This damage results either in the signs being "changed" to plus (as in (b)) or in the signs being "changed" to minus; it is not known which. The damage, when it occurs, is always the same. Find a test of the hypothesis that the probability is zero.

CHAPTER 6

Statistical Estimation

Using the notation of Definition 4.3.3, we now assume that $\mathscr{X} = \mathbb{R}^k$ where k is a given integer. In classical problems, only a finite number of parameters are to be estimated. The following notions are elementary, but more important applications will be developed in the exercises and in Chapter 10. The reader will note that some properties, usually introduced in estimation theory, were introduced in previous chapters when mathematically convenient. For instance, the properties of estimators as statistics were already studied. In Section 1.4 we showed that if a prior distribution is given then there exists a posterior distribution which can give a strategy to estimate the parameter. The Cramer–Rao inequality was established in Section 3.2. Also, one should note that completeness is a fundamental notion for statistical estimation. The presence of nuisance parameters makes constructing confidence regions more difficult. Finally, in general, there is no simple relationship between an estimate of θ and that of a function of θ, if both estimates are to have optimal properties.

1. Unbiased Estimators

Let $[\Omega, \mathscr{A}; P_\theta, \theta \in \Theta]$ be a statistical space and let f be a mapping of Θ in $\mathbb{R}^k$. We have seen in Definition 4.3.3 that an estimator of f is a statistic, taking values in $(\mathbb{R}^k, \mathscr{B}_{\mathbb{R}^k})$. One would like to impose the condition,

$$X(\omega) \in f(\Theta), \qquad \mathscr{P}\text{-a.e.};$$

i.e., the statistic almost always takes on possible values of $f(\Theta)$. But, in some applications, it is troublesome to handle. An example of this occurs when an estimator of a variance takes on negative values.

Definition 1 Let $[\Omega, \mathscr{A}; P_\theta, \theta \in \Theta]$ be a statistical space, f a mapping from Θ in $\mathbb{R}^k$ and X an estimator of f. If X is integrable and has image f we say that X is an *unbiased estimator* of X.

If the statistical space is complete, there exists at most one unbiased estimator of a given function f. If the statistical space is not complete, all unbiased estimators are obtained by adding a statistic which has zero expectation to an unbiased estimator, if one exists.

Definition 2 For every integer n, let X_n be an estimator of f on the sample space $[\Omega, \mathscr{A}; P_\theta, \theta \in \Theta]^n$. X_n is an *asymptotically unbiased estimator* of f if

$$\forall \theta \in \Theta, \qquad \beta_{X_n}(\theta) \underset{n \to \infty}{\rightarrow} f(\theta).$$

If, for every $\theta \in \Theta$, and for a given notion of probabilistic convergence we have $X_n \to f(\theta)$, then we say that X_n is a *consistent estimator* with respect to this notion of convergence.

Theorem 1 *Let X be an unbiased estimate of f in the statistical space $[\Omega, \mathscr{A}; P_\theta, \theta \in \Theta]$. The estimator of f which is defined on the statistical space $[\Omega, \mathscr{A}; P_\theta, \theta \in \Theta]^n$ by*

$$X_n(\omega_1, \ldots, \omega_n) = \frac{1}{n} \sum_{j=1}^{n} X(\omega_j) \qquad (\omega_j \in \Omega, j = 1, \ldots, n)$$

is an unbiased estimator of f which is almost surely consistent.

This theorem is an application of the linearity of mathematical expectation and the strong law of large numbers.

Remark If $f(\Theta)$ is a convex set which contains $X(\Omega)$, then for every n, $f(\Theta)$ contains $X_n(\Omega)$.

Theorem 2 *Let $\mathscr{B}$ be a sufficient subfield on the statistical space $[\Omega, \mathscr{A}; P_\theta, \theta \in \Theta]$ and let X be an unbiased estimator of f. Then the projection of X on $\mathscr{B}$ is also an unbiased estimator of f.*

This theorem results directly from Definition 2.3.1.

In some cases, Theorems 1 and 2 furnish a useful procedure for finding interesting unbiased estimators on a sample. Let $\mathscr{B}$ be a sufficient subfield for the sample $[\Omega, \mathscr{A}; P_\theta, \theta \in \Theta]^n$, then $E(X_n | \mathscr{B})$ is an unbiased, $\mathscr{B}$-measurable estimator of f, which may be difficult to find directly.

Remark If $f(\Theta)$ is a convex set and contains $X(\Omega)$ and if there exists a regular determination of the conditional probability given $\mathscr{B}$, then $f(\Theta)$ also contains $Z(\Omega)$, where

$$Z(\omega) = \int_\Omega X \, dP_{\mathscr{B}}^{\omega} = E(X | \mathscr{B}).$$

Finally, we say that an estimator is sufficient, sufficient with respect to a nuisance parameter, free, and free with respect to a nuisance parameter, if the estimator has these properties when considered a statistic.

2. Optimal Estimators

Using Theorem 4.5.1, we can, generally, restrict ourselves to determining strategies (i.e., estimators) which are optimal among deterministic strategies. For example, if a prior distribution Q is given, we obtain (Section 1.3) a posterior distribution of θ and then an estimator of f defined by

$$E_\pi(f(\theta) | \omega),$$

where the distribution π is defined by (P_θ, Q) on $(\Omega \times \Theta, \mathscr{A} \otimes \mathscr{F})$.

Furthermore, using the Jensen inequality (Theorem A.4.9), we easily prove the following theorem which shows that if there exists a sufficient subfield $\mathscr{B}$ we can restrict ourselves to $\mathscr{B}$-measurable estimators.

Theorem 1 *Let $[\Omega, \mathscr{A}; P_\theta, \theta \in \Theta]$ be a statistical space, f a mapping from Θ into $\mathbb{R}^k$, $\mathscr{B}$ a sufficient subfield, X an integrable estimator of f and let W be a loss function on $\Theta \times \mathbb{R}^k$, which is continuous and convex for every θ. Then we have*

$$E(X | \mathscr{B}) \overset{W}{\geqslant} X.$$

Definition 1 The estimator X of f is a *minimum variance* estimator if X is an unbiased estimator having a covariance matrix which is less, for every θ, than the covariance matrix of any other unbiased estimator of X. (The quasi-ordering on the symmetric matrices is that of the corresponding quadratic forms.)

Note that the optimum is sought among unbiased tests and that a biased estimator can exist having a covariance less than that of X.

Theorem 2 *Let $[\Omega, \mathscr{A}; P_\theta, \theta \in \Theta]$ be a statistical space, $\mathscr{B}$ a sufficient and complete subfield and f a mapping of Θ into $\mathbb{R}^k$. If there exists an unbiased estimator of f having a covariance matrix, then there exists only one ($\mathscr{P}$-a.e.) unbiased and $\mathscr{B}$-measurable estimator of f which has minimum variance.*

Proof The projection on $\mathscr{B}$ of an unbiased estimator of f is an unbiased $\mathscr{B}$-measurable estimator of f. Moreover, since $\mathscr{B}$ is complete, every unbiased estimator of f has the same projection on $\mathscr{B}$. Theorem 1 applied, for every u belonging to $\mathbb{R}^k$, to the loss function

$$\theta \in \Theta, \quad x \in \mathbb{R}^k, \qquad W_u(\theta, x) = \langle u | x - f(\theta) \rangle^2$$

yields the result. ∎

When one has a $\mathscr{P}$-minimum sufficient subfield which is not complete, finding a minimum variance estimator is difficult (see, e.g., Section 10.5 for the case of exponential statistical spaces). In some cases, the following theorem will be useful.

Theorem 3 *Let $[\Omega, \mathscr{A}; P_\theta, \theta \in \Theta]$ be a statistical space and X a statistic, taking values in $(\mathbb{R}^k, \mathscr{B}_{\mathbb{R}^k})$, which has a covariance matrix. Then X is a minimum variance estimator of its image if and only if, for every θ, every component of X is uncorrelated with any real statistic which has zero expectation and a second-order moment.*

Proof Let Z be a statistic taking values on $(\mathbb{R}^k, \mathscr{B}_{\mathbb{R}^k})$ which has a covariance matrix and zero expectation; i.e., every component of Z has zero expectation and a moment of order 2. For every real λ, the statistic

$$X' = X + \lambda Z$$

has the same image as X and we have

$$\forall x \in \mathbb{R}^k, \qquad {}^t x \Lambda_{X'}^\theta x = E_{P_\theta}(\langle x | X' \rangle^2).$$

Then we obtain

$${}^t x \Lambda_{X'}^\theta x = {}^t x \Lambda_X^\theta x + 2\lambda {}^t x \Sigma_{XZ}^\theta x + \lambda^2 {}^t x \Lambda_Z^\theta x.$$

where Σ_{XZ}^θ is the covariance (matrix) between X and Z (see Section 7.4) and Λ_X^θ, Λ_Z^θ are the covariance matrices of X and Z respectively.

To show sufficiency let X' be another unbiased estimator. Then $X - X'$ has zero expectation. But, by hypothesis, X and $X - X'$ are uncorrelated. Thus we have

$$\forall x \in \mathbb{R}^k, \qquad {}^t x \Lambda_{X'}^{\theta} x = {}^t x \Lambda_X^{\theta} x + \lambda^2\, {}^t x \Lambda_{X-X'}^{\theta} x,$$

and the variance of X is minimum.

Conversely, if the variance of X is minimum we have

$$\forall \lambda \in \mathbb{R}, \quad \theta \in \Theta, \quad x \in \mathbb{R}^k, \qquad 2\lambda\, {}^t x \Sigma_{XZ}^{\theta} x + \lambda^2\, {}^t x \Lambda_Z^{\theta} x \geqslant 0,$$

and then $\Sigma_{XZ}^{\theta} = 0$. ∎

Remark 1 We say that an estimator is *efficient* if it is unbiased and if using Theorem 3.2.5, we have

$$\Lambda_X = \Delta \mathscr{I}^{-1}\, {}^t\Delta.$$

This is clearly a minimum variance estimator. However, the corollary of the theorem shows that efficiency is possible only in special cases, so that this procedure is not very useful.

Remark 2 Suppose that the statistical space is dominated and that for every ω the likelihood function $\mathscr{L}(\theta, \omega)$ is maximized for the unique value $\hat{\theta}(\omega)$. Moreover, if the function $\hat{\theta}(\omega)$ is measurable then $\hat{\theta}$ is a statistic which is called the *maximum likelihood statistic*. Then $f(\hat{\theta}(\omega))$ is called a *maximum likelihood estimator* of $f(\theta)$ and has many interesting properties (see e.g., Dugué [14, p. 140]).

3. Construction of Confidence Regions

Let $[\Omega, \mathscr{A}; P_\theta, \theta \in \Theta]$ be a statistical space and f be a mapping of Θ in $\mathbb{R}^k$. Using Definition 4.3.4, we know that a *set estimator* of f is a mapping d from Ω in $\mathscr{B}_{\mathbb{R}^k}$ such that

$$\forall x \in \mathbb{R}^k, \qquad d^{-1}(x) = \{\omega; x \in d(\omega)\} \in \mathscr{A}.$$

Similarly, as in the case of an estimator, we could introduce the condition

$$f(\Theta) \supset d(\omega), \qquad \forall \omega \in \Omega,$$

which is convenient. However, for the same reason as in Section 1, we do not systematically make this assumption, and do not require that $d(\omega)$ be nonempty for all ω.

Definition 1 Let $[\Omega, \mathscr{A}; P_\theta, \theta \in \Theta]$ be a statistical space, f a mapping from Θ into $\mathbb{R}^k$ and d a set estimator of f. We call the *level of significance* of this set estimator the real number α_d defined by

$$\alpha_d = 1 - \inf[P_\theta(d^{-1}(f(\theta))) | \theta \in \Theta]$$

and we say that this set estimator is *free* if

$$\forall \theta \in \Theta, \qquad P_\theta(d^{-1}(f(\theta))) = 1 - \alpha_d.$$

Theorem 1 *Let D be a strong set estimator having the level of significance α (and be free). For every prior distribution Q we have*

$$\pi(\Delta) \geqslant 1 - \alpha \qquad (\text{or } \pi(\Delta) = 1 - \alpha),$$

where

$$\Delta = \{(\theta, \omega) : (f(\theta), \omega) \in D\},$$

and π is the distribution on $\Omega \times \Theta$ generated by (P_θ, Q).

Proof Recall that a strong set estimator D is a subset of $\Omega \times \mathbb{R}^k$, belonging to $\mathscr{A} \otimes \mathscr{B}_{\mathbb{R}^k}$. For every ω belonging to Ω, let $d(\omega)$ be the Borel set defined by the values of x such that (ω, x) belongs to D. Then the mapping $\omega \to d(\omega)$ is the set estimator deduced from D and we have

$$\pi(\Delta) = \int_\Theta \pi(\Delta | \theta)\, dQ(\theta) = \int_\Theta P_\theta(d^{-1}(f(\theta)))\, dQ(\theta),$$

whence the proof is completed by using Definition 1. ∎

Remark 1 Note the connection between this theorem and the concept of Fisher's fiducial probability [19].

Remark 2 If f is not a one-to-one mapping, in many cases, by performing a transformation on the parameter, we can reduce the problem to the case,

$$\theta = (x, \theta'), \quad x \in \mathscr{X} \subset \mathbb{R}^k, \quad \theta' \in \Theta'; \qquad \Theta = \mathscr{X} \times \Theta'; \qquad f(\theta) = x,$$

where x is the principal parameter and θ' the nuisance parameter. Then we see that to determine a free set estimator of x, we must find, for every x belonging to $\mathscr{X}$, an event which is free with respect to θ'.

The following theorem shows that we can deduce a set estimator from a family of deterministic tests having the same level of significance. For every ω, we take as $d(\omega)$ the subset of values of $f(\theta)$ such that if we test each of these values against another hypothesis, we accept these values when ω is given.

Theorem 2 *Let d be a set estimator of f having level of significance α*

(and be free). Then, for every $\theta \in \Theta$, the deterministic test Φ of $f^{-1}(f(\theta))$ against another hypothesis, defined by the critical region $\complement d^{-1}(f(\theta))$ has a level of significance less than α (and is free). Conversely if for every $x \in f(\Theta)$ there exists a deterministic test of $f^{-1}(\{x\})$ against another hypothesis, having level of significance α (and is free), then the mapping

$$\omega \rightarrow d(\omega) = \{x \in f(\Theta) : \omega \in \complement C_x\}$$

where C_x is the critical region of the previous test for the value x, is a set estimator having level of significance α (and is free).

Proof We have

$$\forall \theta' \in \Theta, \qquad \beta_\Phi(\theta') = 1 - P_{\theta'}(d^{-1}(f(\theta))),$$

and in particular

$$\forall \theta' \in f^{-1}(f(\theta)), \qquad f(\theta') = f(\theta), \qquad d^{-1}(f(\theta)) = d^{-1}(f(\theta')).$$

Then

$$\forall \theta' \in f^{-1}(f(\theta)), \qquad \beta_\Phi(\theta') = 1 - P_{\theta'}(d^{-1}(f(\theta'))),$$

and finally

$$\alpha_\Phi = \sup(\beta_\Phi(\theta') | f^{-1}(f(\theta))) \leqslant 1 - \inf[P_{\theta'}(d^{-1}(f(\theta'))) | \Theta] = \alpha.$$

Furthermore, if d is free we have

$$\forall \theta' \in f^{-1}(f(\theta)), \qquad \beta_\Phi(\theta') = 1 - P_{\theta'}(d^{-1}(f(\theta'))) = \alpha$$

and then Φ is free.

Conversely, if d is the set estimator given in the theorem, we have

$$d^{-1}(x) = \complement C_x, \qquad \forall x \in f(\Theta).$$

But by hypothesis

$$\forall \theta \in f^{-1}(\{x\}), \qquad P_\theta(C_x) \leqslant \alpha \qquad (\text{or } P_\theta(C_x) = \alpha).$$

Then in particular

$$\forall \theta \in \Theta, \qquad 1 - P_\theta(d^{-1}(f(\theta))) \leqslant \alpha \qquad (\text{or } P_\theta(d^{-1}(f(\theta))) = 1 - \alpha). \qquad \blacksquare$$

4. Optimal Set Estimators

Here we look for the best set estimators among all those having a given level of confidence. Let μ be a positive measure on $(\mathbb{R}^k, \mathscr{B}_{\mathbb{R}^k})$. The random variable $\mu(d(\omega))$ will be used as a measure of merit of the set estimator d; in most applications we use Lebesgue measure for μ

Definition 1 Let $[\Omega, \mathscr{A}; P_\theta, \theta \in \Theta]$ be a statistical space, f a mapping of Θ into $\mathbb{R}^k$, d a set estimator of f and μ a positive measure on $(\mathbb{R}^k, \mathscr{B}_{\mathbb{R}^k})$. We say that d is *μ-optimal* with respect to a given class $\mathscr{F}$ of set estimators if, for every set estimator d' belonging to $\mathscr{F}$, we have

$$\alpha_d \geqslant \alpha_{d'} \quad \Rightarrow \quad \mu(d'(\omega)) \geqslant \mu(d(\omega)), \qquad \forall \omega \in \Omega.$$

Theorem 1 *Suppose that $f(\Theta)$ is a Borel set of $\mathbb{R}^k$ and that there exists a $\mathscr{A} \otimes \mathscr{B}_{f(\Theta)}$-measurable function $G(\omega, x)$ on $\Omega \times f(\Theta)$ taking values on $(\mathscr{Y}, \mathscr{D})$, such that for every θ the random element $\omega \to G(\omega, f(\theta))$ has the same distribution Q. Then, for every set U belonging to $\mathscr{D}$ the set $G^{-1}(U)$ is a free, strong set estimator of f which has level of significance $1 - Q(U)$.*

The function G is often called a *pivotal* function.

Proof The mapping

$$d_U(\omega) = \{x : G(\omega, x) \in U, x \in f(\Theta)\}$$

is a set estimator and we have

$$\forall \theta \in \Theta, \qquad d_U^{-1}(f(\theta)) = \{\omega : G(\omega, f(\theta)) \in U\}.$$

Then

$$P_\theta(d_U^{-1}(f(\theta)) = P_\theta(\{G(\omega, f(\theta)) \in U\}) = Q(U). \qquad \blacksquare$$

Theorem 2 *Assuming the conditions of Theorem 1, let μ be a measure on $(f(\Theta), \mathscr{B}_{f(\Theta)})$ and μ_ω^G the measure on $(\mathscr{Y}, \mathscr{D})$ which is induced from μ by the mapping $x \to G(\omega, x)$. If there exists a set $U^* \in \mathscr{D}$, such that*

$$Q(U) \geqslant Q(U^*) \quad \Rightarrow \quad \mu_\omega^G(U) \geqslant \mu_\omega^G(U^*), \qquad \forall \omega \in \Omega, \tag{1}$$

then $G^{-1}(U^*)$ is a strong set estimator which is μ-optimal with respect to the family of set estimators obtained by using Theorem 1.

Proof Theorem A.1.1 shows that

$$\forall U \in \mathscr{D}, \quad \forall \omega \in \Omega, \qquad \mu(d_U(\omega)) = \mu_\omega^G(U)$$

and condition (1) is then equivalent to the condition of Definition 1. ∎

Note that if the set U^* exists, we can find it by using the Neyman–Pearson lemma applied to (1). See, for example Exercises 7 and 11.

Exercises

1. Consider the statistical space, $[\mathbb{R}, \mathscr{B}_{\mathbb{R}}; N(m, \sigma^2), m \in \mathbb{R}, \sigma \in \mathbb{R}^+]^n$. Find an optimal estimator of σ^{2k}, where k is a given positive real number.

2. Let $\{X_n\}$ be a sequence of random variables such that

$$E(X_n) = m, \qquad \sigma^2(X_n) = \sigma^2/n,$$

where m and σ^2 are given real numbers.

(a) Show that

$$X_n \xrightarrow{P} m \qquad \text{as} \quad n \to \infty.$$

(b) Assume furthermore that when $n \to \infty$ the distribution of $n^{1/2}(X_n - m)$ has a limit distribution having zero expectation and variance σ^2. Let f be a real function. Show that, in general, the distribution of the random variable

$$Y_n = n^{1/2}[f(X_n) - f(m)]$$

has a limit; compute this limit, its expectation and its variance.

(c) Applying the results of (a) and (b) to the theory of statistical estimation, find the assumptions which are to be made on a given function f, such that we can deduce from an unbiased and convergent estimator of θ, a convergent and asymptotically unbiased estimator of $f(\theta)$.

3. Consider the following statistical space $[\mathbb{R}, \mathscr{B}_{\mathbb{R}}, \mathscr{P}]^n$ and let T be a real Borel function on $\mathbb{R}^n$. We want to estimate

$$\theta = \int_{\mathbb{R}^n} T\, dP^n \qquad (P \in \mathscr{P}).$$

(a) Compare the estimator T and the estimator

$$\tilde{T}(x_1, \ldots, x_n) = \frac{1}{n!} \sum_{\pi} T(\pi(x_1, \ldots, x_n)),$$

where π runs through all possible permutations.

(b) Apply (a) to the following case: P is the Bernoulli distribution defined by $P(\{X = 1\}) = p$ and T is equal to zero if all variables $x_1, \ldots, x_n$ are zero, and if not, T is equal to the rank of the first variable which is equal to 1. Compute θ and the statistic $\tilde{T}$.

4. Let v be an observation from a binomial distribution with p the unknown parameter.

(a) Let α be a given positive integer less than n, show that the statistic $[v]^\alpha/[n]^\alpha$ (where $[x]^\alpha = x(x-1)\cdots(x-\alpha+1)$) is an unbiased estimator of p^α.

(b) Find all the functions of p which admit an unbiased estimator.

(c) Find an estimator for $p^\alpha(1-p)^\beta$ (α and β given) and discuss the case where β is negative.

5. Consider the statistical space $[\mathbb{R}, \mathscr{B}_{\mathbb{R}}, N(m, 1), m \in \mathbb{R}]^n$ and let a and $b (a < b)$ be two given real numbers. We want to estimate $\theta = f(m) = \int_a^b dP$. Let n_1 be the number of observation which belong to $[a, b]$. Show that n_1/n is an unbiased estimator of θ and deduce from it a minimum variance unbiased estimator of θ. Comment.

6. Consider the statistical space $[\mathbb{R}, \mathscr{B}_{\mathbb{R}}; N(m, \sigma^2), m \in \mathbb{R}]$ for given σ and the polynomials

$$Q_s(u) = \sum_{j=0}^{s} \frac{2s!}{(2j)!(s-j)!}\left(-\frac{1}{2}\right)^{s-j} u^j,$$

$$Q'_s(u) = \sum_{j=0}^{s} \frac{(2s+1)!}{(2j+1)!(s-j)!}\left(-\frac{1}{2}\right)^{s-j} u^j, \qquad s = 0, 1, \ldots.$$

Show that, for every positive integer s, $\sigma^{2s}Q_s(x^2/\sigma^2)$ and $\sigma^{2s}xQ'_s(x^2/\sigma^2)$ are unbiased estimators of m^{2s} and m^{2s+1}, respectively. State and show a similar result when considering a sample of size n from the above statistical space.

7. Using notation of Exercise 6, let α be a given number in $]0, 1[$. Show that the interval

$$(x + \sigma N^{-1}(\alpha/2), x - \sigma N^{-1}(\alpha/2))$$

is a free confidence interval of m, which has level of significance α, whose length is minimum among all set estimators which depend only on $x - m$ and have level of significance less than or equal to α.

8. Consider the statistical space $[\mathbb{R}^+, \mathscr{B}_{\mathbb{R}^+}; \Gamma(a, \lambda), \lambda \in \mathbb{R}^+]$ (see Section 7.1) where a is a given real nonnegative number. Show that for every given real number s which is greater than or equal to $-a$,

$$\frac{\Gamma(a)}{\Gamma(a+s)} x^s$$

is an unbiased estimator of λ^s.

9. For the space in Exercise 8, let α be a given number in $]0, 1[$. Find the free confidence interval for λ^s which has level of significance α and whose length is minimum among all set estimators which have level of significance α and depend only on λx.

10. Consider the statistical space $[\mathbb{R}, \mathscr{B}_{\mathbb{R}}; N(m, \sigma^2), m \in \mathbb{R}, \sigma \in \mathbb{R}^+]^n$, let

k be a nonnegative integer and γ a real number ($\gamma > 1 - n$). Show that the statistics,

$$Q_{k\gamma}(\bar{x}, s) = \left(\frac{n}{2}\right)^{\gamma/2} \Gamma\left(\frac{n-1}{2}\right) \times \sum_{j=0}^{k} \frac{(2k)!(-4)^{j-k}}{(2j)!(k-j)!\Gamma\left(\frac{n+\gamma-1}{2} + k - j\right)} \bar{x}^{2j} s^{k-j+(\gamma/2)}$$

$$Q'_{k\gamma}(\bar{x}, s) = \left(\frac{n}{2}\right)^{\gamma/2} \Gamma\left(\frac{n-1}{2}\right)\bar{x} \times \sum_{j=0}^{k} \frac{(2k+1)!(-4)^{j-k}}{(2j+1)!(k-j)!\Gamma\left(\frac{n+\gamma-1}{2} + k - j\right)} \bar{x}^{2j} s^{k-j+(\gamma/2)},$$

where

$$\bar{x} = \frac{1}{n}\sum_{1}^{n} x_i \quad \text{and} \quad s = \frac{1}{n}\sum_{1}^{n} (x_i - \bar{x})^2,$$

are unbiased estimators of $m^{2k}\sigma^{\gamma}$ and $m^{2k+1}\sigma^{\gamma}$, respectively.

11. For the space in Exercise 10, let α be a given number in $]0, 1[$. Find a constant $K_n(\alpha)$ such that the convex set defined by

$$(m - \bar{x})^2 + s \leqslant \frac{n+1}{n} \sigma^2 \log\left(\frac{s}{\sigma^2} K_n(\alpha)\right)$$

is a free confidence region for (m, σ) having level of significance α and whose surface area is minimum among the estimators which have level of significance less than or equal to α and which are functions only of $(\bar{x} - m)/\sigma$ and s/σ^2.

12. Let U be a random variable having the distribution $N(m, \sigma^2)$. Find $Q(m, \sigma)$, the distribution of e^U and compute its moments. Consider the following statistical space

$$[\mathbb{R}, \mathscr{B}_{\mathbb{R}}; Q(m, \sigma), m \in \mathbb{R}, \sigma \in \mathbb{R}^+]^n.$$

(a) Find a sufficient statistic.

(b) Find the maximum likelihood estimator of the expectation of $Q(m, \sigma)$. Show that this estimator is biased but asymptotically unbiased and efficient.

(c) Show that there exists a minimum variance unbiased estimator and compute it.

13. Show that in statistical estimation problems, without a nuisance parameter, if there exists a sufficient statistic and a maximum likelihood estimator V, then V depends only on T.

14. Let P be the exponential distribution on $\mathbb{R}^+$ and let P_θ be the distribution which is deduced from P by the translation θ. Find and study the maximum likelihood estimator of θ for the statistical space

$$[\mathbb{R}, \mathscr{B}_{\mathbb{R}}; P_\theta, \theta \in \mathbb{R}]^n.$$

15. Let X be a random variable having the $\mathscr{B}(n; p, q)$ distribution. Find the limit distribution of the random variable

$$Y = 2n^{1/2}[\arcsin (X/n)^{1/2} - \arcsin p^{1/2}]$$

and deduce from this result a confidence interval for p having level of significance α. Compare it with the confidence interval deduced from the normal approximation.

16. Suppose that we have a sample on a multinomial random variable, where p is the vector of probabilities.

(a) Find the maximum likelihood estimate of p and compute its variance.

(b) Choose a set of linearly independent parameters and compute the information matrix. Compare the result with that of (a).

CHAPTER 7

The Multivariate Normal Distribution

This chapter deals with probability; it introduces distributions used most frequently in classical statistical problems based on the normal distribution and gives relationships among these distributions.

1. Some Useful Distributions

The reader should refer to elementary probability texts for details on the elementary distributions, such as the *binomial* distribution $\mathscr{B}(n; p, q)$ whose generating function is $(q + pu)^n$ where $p + q = 1$, the *Poisson* distribution $\mathscr{P}(\lambda)$ whose probability for integral k is $e^{-\lambda}\lambda^k/k!$ and the *normal* distribution $N(m, \sigma^2)$ whose density is

$$\frac{1}{\sigma(2\pi)^{1/2}} \exp\left\{-\frac{1}{2\sigma^2}(x - m)^2\right\}.$$

The gamma and beta distributions to be discussed now will also prove to be useful in statistics.

Definition 1 (*Gamma Distribution*) For positive parameters a and λ we write $\Gamma(a, \lambda)$ for the distribution on $(\mathbb{R}^+, \mathscr{B}_{\mathbb{R}^+})$ defined by the probability density function,

$$\frac{\lambda^{-a}}{\Gamma(a)} e^{-x/\lambda} x^{a-1}.$$

The reader can verify easily that the characteristic function of this distribution is $(1 - it\lambda)^{-a}$ and that the moment of order k is

$$\lambda^k \frac{\Gamma(a+k)}{\Gamma(a)} = \lambda^k(a+k-1)\cdots a.$$

If the random variable X has the distribution $\Gamma(a, \lambda)$, then the random variable cX has the distribution $\Gamma(a, c\lambda)$. If n is a positive integer, $\Gamma(n/2, 2)$ is called the χ^2-distribution with n degrees of freedom.

Definition 2 (*Beta Distributions*) For all positive parameters a, b, we write $\beta(a, b)$ for the distribution on $(\mathbb{R}^+, \mathscr{B}_{\mathbb{R}^+})$ defined by the probability density function,

$$\frac{\Gamma(a+b)}{\Gamma(a)\Gamma(b)} x^{a-1}(1+x)^{-a-b}. \tag{1}$$

We write $\beta'(a, b)$ for the distribution on $([0, 1], \mathscr{B}_{[0,1]})$ defined by the probability density function,

$$\frac{\Gamma(a+b)}{\Gamma(a)\Gamma(b)} y^{a-1}(1-y)^{b-1}. \tag{2}$$

Expressions (2) and (1) are sometimes referred to as *beta* densities of the *first and second kind*, respectively.†

Definition 3 (*Noncentral Gamma and Beta Distributions*) For all positive parameters a, b, γ, λ we define the noncentral distributions $\Gamma(a, \gamma, \lambda)$, $\beta(a, b, \gamma)$, and $\beta'(a, b, \gamma)$ by

$$\Gamma(a, \gamma, \lambda) = e^{-\gamma/\lambda} \sum_{m=0}^{\infty} \frac{(\gamma/\lambda)^m}{m!} \Gamma(a+m, \lambda)$$

$$\beta(a, b, \gamma) = e^{-\gamma} \sum_{m=0}^{\infty} \frac{\gamma^m}{m!} \beta(a+m, b)$$

$$\beta'(a, b, \gamma) = e^{-\gamma} \sum_{m=0}^{\infty} \frac{\gamma^m}{m!} \beta'(a+m, b).$$

If the random variable X has the $\Gamma(a, \gamma, \lambda)$ distribution, then cX has the $\Gamma(a, c\gamma, c\lambda)$ distribution. The parameter γ is called the noncentrality param-

† Although the usual American terminology for the beta density is (2) we shall generally refer to (1) as the beta density. (Ed.)

eter. One can show easily that the characteristic function of the $\Gamma(a, \gamma, \lambda)$ distribution is

$$(1 - it\lambda)^{-a} \exp\left\{\frac{it\gamma}{1 - it\lambda}\right\}$$

and thus the existence of the reproductive convolution formula

$$\Gamma(a, \gamma, \lambda) * \Gamma(b, \gamma', \lambda) = \Gamma(a + b, \gamma + \gamma', \lambda).$$

The reader should show as an exercise the following:

Theorem 1 *Let X and Y be two independent random variables having $\Gamma(a, \gamma, \lambda)$ and $\Gamma(b, \lambda)$ distributions respectively, then the random variables*

$$U = X + Y, \qquad V = X/Y, \qquad W = X/(X + Y)$$

have $\Gamma(a + b, \gamma, \lambda)$, $\beta(a, b, \gamma/\lambda)$, $\beta'(a, b, \gamma/\lambda)$ distributions respectively. If $\gamma = 0$, V and W are each independent of U.

Using this theorem and Exercise 4 the following theorems, which will be useful in applications, can be established.

Theorem 2 *Let $X_1, \ldots, X_n$ be n independent random variables distributed as $N(m_i, \sigma^2)$, $i = 1, \ldots, n$. Then the random variable $Y = X_1^2 + \cdots + X_n^2$ has the $\Gamma(n/2, m, 2\sigma^2)$ distribution where $m = m_1^2 + \cdots + m_n^2$.*

Theorem 3 *Let $X_1, \ldots, X_n$; $Y_1, \ldots, Y_p$ be independent normal random variables. If all these random variables have the same variance σ^2, and if the random variables $Y_1, \ldots, Y_p$ have zero expectations, then the random variable*

$$T = \frac{X_1^2 + \cdots + X_n^2}{Y_1^2 + \cdots + Y_p^2}$$

has the $\beta(n/2, p/2, m/2\sigma^2)$ distribution where $m = [E(X_1)]^2 + \cdots + [E(X_n)]^2$.

2. Multivariate Normal Distributions

Lemma 1 *Let M and Λ be two symmetric matrices of order n and let r be the rank of Λ which is assumed to be nonnegative definite. Then there exists at least one matrix A of order (n, r) such that*

$$\Lambda = A\,{}^{t}A, \qquad {}^{t}AMA = K,$$

where K is a diagonal matrix having the same positive eigenvalues as $M\Lambda$.

Proof There exists at least one orthogonal matrix C such that

$$C\Lambda\,{}^{t}C = \begin{pmatrix} \lambda_1 & & & & 0 \\ & \cdot & & & \\ & & \cdot & & \\ & & & \lambda_r & \\ 0 & & & & 0 \end{pmatrix}$$

where $\lambda_1, \ldots, \lambda_r$ are the positive eigenvalues of Λ. Let K_0 be the matrix

$$\begin{pmatrix} (\lambda_1)^{1/2} & & & 0 \\ & \cdot & & \\ & & \cdot & \\ 0 & & & (\lambda_r)^{1/2} \end{pmatrix}$$

and let C_0 be the matrix of the r first rows of C. Now let C' be an orthogonal matrix of order r. Then the matrix

$$A = {}^{t}C_0 K_0 C'$$

is such that $\Lambda = A\,{}^{t}A$, since we have

$$A\,{}^{t}A = {}^{t}C_0 K_0^2 C_0 = {}^{t}C\begin{pmatrix} K_0^2 & 0 \\ 0 & 0 \end{pmatrix}C = \Lambda.$$

Consider the symmetric matrix $K_0 C_0 M\,{}^{t}C_0 K_0$. There exists an orthogonal matrix C_1 of order r such that the matrix

$${}^{t}C_1 K_0 C_0 M\,{}^{t}C_0 K_0 C_1$$

is diagonal. The proof will be completed if we show that the matrix ${}^{t}C_0 K_0 C_1$ satisfies the conditions of the lemma; i.e., if $M\Lambda$ and $K_0 C_0 M\,{}^{t}C_0 K_0$ have the same positive eigenvalues. But using the corollary of Theorem 8.3.7, we see that $K_0 C_0 M\,{}^{t}C_0 K_0$ has the same nonzero eigenvalues as $M\,{}^{t}C_0 K_0^2 C_0 = M\Lambda$. ∎

Theorem 1 *Let m be a vector of $\mathbb{R}^n$ and let Λ be a symmetric nonnegative definite matrix of order n. There exists a distribution on $(\mathbb{R}^n, \mathscr{B}_{\mathbb{R}^n})$ which is called the normal distribution and has the following characteristic function*

$$\exp\{i\,{}^{t}tm - \tfrac{1}{2}\,{}^{t}t\Lambda t\}, \qquad t \in \mathbb{R}^n.$$

We designate this distribution as $N(m, \Lambda)$.

Theorem 2 *The random vector X having the distribution $N(m, \Lambda)$ is said to be a normal vector. The expectation of X is m and its covariance matrix is Λ.*

Theorem 3 *Let X be a random vector having the distribution $N(m, \Lambda)$ and let A be a matrix which has as many columns as X has components. Then the distribution of AX is*

$$N(Am, A\Lambda\,{}^tA).$$

Theorem 4 *Let X be a normal vector with a diagonal covariance matrix; then the components of X are independent random variables.*

Theorem 5 *Let X be a random vector taking values on $(\mathbb{R}^n, \mathscr{B}_{\mathbb{R}^n})$ which has the distribution $N(m, \Lambda)$. Let r be the rank of Λ; then there exists at least one matrix D of order (n, r) such that*

$$X = m + DY,$$

where Y is a random vector having the distribution $N(0, \mathbb{1}_r)$.

Theorem 6 *If the matrix Λ is nonsingular, then the distribution $N(m, \Lambda)$ has the probability density function,*

$$(2\pi)^{-n/2}(\det \Lambda)^{-1/2} \exp\{-\tfrac{1}{2}\,{}^t(x - m)\Lambda^{-1}(x - m)\}.$$

Proof of Theorems 1 *to* 6 Let $X_1, \ldots, X_n$ be n independent random variables with distributions $N(m_1, \sigma_1^2), \ldots, N(m_n, \sigma_n^2)$, respectively. Write

$$X = \begin{pmatrix} X_1 \\ \vdots \\ X_n \end{pmatrix}, \qquad m = \begin{pmatrix} m_1 \\ \vdots \\ m_n \end{pmatrix}, \qquad K = \begin{pmatrix} \sigma_1^2 & & 0 \\ & \ddots & \\ 0 & & \sigma_n^2 \end{pmatrix}.$$

It is easy to see that the characteristic function of X is

$$\forall t \in \mathbb{R}^n, \qquad \varphi_X(t) = \exp\{i\,{}^ttm - \tfrac{1}{2}\,{}^ttKt\}$$

and Theorem 4 is demonstrated. Now let Λ be a symmetric nonnegative definite matrix of order n. Then, if r is the rank of Λ, using the lemma we know that there exists a matrix A of order (n, r) such that $\Lambda = A\,{}^tA$. Let Y be a random vector having the distribution $N(0, \mathbb{1}_r)$; using Theorem 4, its characteristic function is equal to $\exp\{-\frac{1}{2}\|t\|^2\}$. If we now consider the random vector $X = m + AY$, we have

$$\varphi_X(t) = \exp[i\,{}^t tm]\varphi_Y({}^t At).$$

Then

$$\varphi_X(t) = \exp\{i\,{}^t tm - \tfrac{1}{2}{}^t tA\,{}^t At\} = \exp\{i\,{}^t tm - \tfrac{1}{2}{}^t t\Lambda t\}$$

and Theorems 1 and 5 are demonstrated. Theorem 2 is easily deduced from Theorem 5, and Theorem 3 from Theorem 1.

Now, we consider a special case of Theorem 5 where $r = n$. Then the random vector Y has probability density function

$$(2\pi)^{-r/2} \exp\{-\tfrac{1}{2}\|y\|^2\}$$

and DY has probability density function

$$(2\pi)^{-r/2} \exp\{-\tfrac{1}{2}{}^t y\,{}^t D^{-1} D^{-1} y\} \det D^{-1}.$$

Theorem 6 is demonstrated by using

$$\Lambda = D\,{}^t D, \qquad \det \Lambda = (\det D)^2 \qquad \blacksquare$$

3. Quadratic Forms of Normal Vectors

Theorem 1 *Let M be a symmetric matrix of order n and let X be an n-dimensional random vector having distribution $N(m, \Lambda)$. Then the characteristic function of the random variable*

$$Z = {}^t XMX$$

is equal to

$$[\det(\mathbb{1}_n - 2itM\Lambda)]^{-1/2} \exp\{it\,{}^t m(\mathbb{1}_n - 2itM\Lambda)^{-1} Mm\}.$$

If the matrix $M\Lambda$ has q positive eigenvalues which are all equal to s, the others being zero, and if

$${}^t mM\Lambda M\Lambda Mm = s\,{}^t mM\Lambda Mm,$$

then the distribution of the random variable

$$Z - {}^t m\left(M - \frac{M\Lambda M}{s}\right)m \quad \text{is} \quad \Gamma\left(\frac{q}{2}, {}^t m\frac{M\Lambda M}{s}m, 2s\right).$$

Proof Let A be the matrix obtained by applying Lemma 2.1. We have

$$\Lambda = A\,{}^{t}A, \qquad {}^{t}AMA = \begin{pmatrix} s_1 & & 0 \\ & \ddots & \\ 0 & & s_r \end{pmatrix} = K.$$

Let Y be a random vector having distribution $N(0, \mathbb{1}_r)$. Then

$$X = m + AY.$$

Then by letting

$${}^{t}u = (u_1, \ldots, u_r) = {}^{t}mMA$$

we have

$$Z = {}^{t}mMm + 2\,{}^{t}uY + {}^{t}YKY$$

and

$$Z = {}^{t}mMm + \sum_{j=1}^{r} (s_j Y_j^2 + 2u_j Y_j). \tag{1}$$

Using Theorem 1.2 an elementary computation shows that if U is a random variable having the distribution $N(0, 1)$, then for all real numbers u and s, the characteristic function of the random variable

$$sU^2 + 2uU$$

is

$$(1 - 2its)^{-1/2} \exp\left\{ -\frac{2u^2t^2}{1 - 2its} \right\}.$$

Therefore the characteristic function of Z is equal to

$$t \in \mathbb{R}, \qquad \left[\prod_{j=1}^{p} (1 - 2its_j) \right]^{-1/2} \exp\left\{ it\,{}^{t}mMm - 2t^2 \sum_{j=1}^{r} \frac{u_j^2}{1 - 2its_j} \right\}. \tag{2}$$

Using Lemma 2.1, we have

$$\forall t \in \mathbb{R}, \qquad \prod_{j=1}^{r} (1 - 2its_j) = \det(\mathbb{1}_n - 2itM\Lambda).$$

We also have

$$\sum_{j=1}^{r} \frac{u_j^2}{1 - 2its_j} = {}^{t}u(\mathbb{1}_r - 2itK)^{-1}u = {}^{t}mMA(\mathbb{1}_r - 2itK)^{-1}\,{}^{t}AMm. \tag{3}$$

Thus the characteristic function of Z is

$$[\det(\mathbb{1}_n - 2itM\Lambda)]^{-1/2} \exp\{it\,{}^t mBMm\},$$

where the matrix B is

$$\mathbb{1}_n + 2itMA(\mathbb{1}_r - 2itK)^{-1}\,{}^t A.$$

Now we must show that

$$\forall t \in \mathbb{R}, \qquad B[\mathbb{1}_n - 2itM\Lambda] = \mathbb{1}_n,$$

or that

$$\forall t \in \mathbb{R}, \qquad -M\Lambda + MA(\mathbb{1}_r - 2itK)^{-1}\,{}^t A(\mathbb{1}_n - 2itM\Lambda) = 0.$$

But $\Lambda = A\,{}^t A$. Thus we must show that

$${}^t A = (\mathbb{1}_r - 2itK)^{-1}\,{}^t A(\mathbb{1}_n - 2itMA)$$

or that

$$(\mathbb{1}_r - 2itK)\,{}^t A = {}^t A(\mathbb{1}_n - 2itM\Lambda),$$

which can be verified easily, and the first part of the theorem is demonstrated.

Now, if the matrix $M\Lambda$ has q positive eigenvalues equal to s, we see that

$$\forall t \in \mathbb{R}, \qquad (\mathbb{1}_r - 2itK)^{-1} = \mathbb{1}_r + \frac{2itK}{1 - 2its}.$$

Then, using (3) we have

$$\sum_{j=1}^{r} \frac{u_j^2}{1 - 2its_j} = {}^t mM\Lambda Mm + \frac{2it}{1 - 2its}\,{}^t mM\Lambda M\Lambda Mm.$$

That is, the second assumption implies that

$$\forall t \in \mathbb{R}, \qquad \sum_{j=1}^{r} \frac{u_j^2}{1 - 2its_j} = \frac{{}^t mM\Lambda Mm}{1 - 2its}.$$

Then the characteristic function of Z is

$$(1 - 2its)^{-q/2} \exp\left\{it\,{}^t m\left(M - \frac{M\Lambda M}{s}\right)m\right\} \exp\left\{\frac{it\,{}^t m(M\Lambda M/s)m}{1 - 2its}\right\}. \qquad \blacksquare$$

The following follows directly from the theorem.

Corollary 1 *If the matrix Λ is nonsingular, the random variable ${}^t X\Lambda^{-1}X$ has the $\Gamma(n/2, {}^t m\Lambda^{-1}m, 2)$ distribution.*

Corollary 2 *Under the general conditions of Theorem* 1, *we have*

$$E(Z) = {}^t mMm + \mathrm{Tr}(M\Lambda).$$

Proof Using (1) we have

$$E(Z) = {}^t mAm + \sum_{j=1}^{r} s_j.$$

However,

$$\sum_{j=1}^{r} s_j = \mathrm{Tr}(K) = \mathrm{Tr}(M\Lambda). \qquad \blacksquare$$

Theorem 2 (Karl Pearson) *Let X be a discrete random variable taking the values $a_j, j = 1, \ldots, K$ with probabilities $p_j, j = 1, \ldots, K$, respectively. Let $v_j, j = 1, \ldots, K$ be the number of occurrences of the value a_j among n independent trials of X. Then, as n increases to infinity, the distribution of the random variable*

$$\chi^2 = \sum_{j=1}^{K} \frac{(v_j - np_j)^2}{np_j}$$

converges to $\Gamma(\frac{1}{2}(K-1), 2)$.

Proof An elementary computation shows that as n increases to infinity the distribution of the random vector

$$\Delta = \left(\frac{v_1 - np_1}{n^{1/2}}, \ldots, \frac{v_k - np_k}{n^{1/2}}\right)$$

converges to the normal distribution, with zero expectation and covariance matrix $\Lambda = \{\lambda_{ij}\}$, where

$$\lambda_{ii} = p_i(1 - p_i), \qquad i = 1, \ldots, K;$$

$$\lambda_{ij} = -p_i p_j, \qquad i \neq j = 1, \ldots, K.$$

Now one can apply Theorem 1 with $m = 0$ and diagonal matrix M with $1/p_j$ as diagonal elements. We verify that the matrix $M\Lambda$ has one zero eigenvalue and $K - 1$ others equal to 1. $\blacksquare$

4. Stochastic Dependence among Normal Vectors

In this Section, X will be a p-dimensional random vector and Y a q-dimensional random vector, each having a covariance matrix. We denote by

Z the random vector $\begin{pmatrix} X \\ Y \end{pmatrix}$ and by Λ_X, Λ_Y, Λ_Z the covariance matrix of X, Y, Z respectively. The matrix

$$\Sigma_{XY} = E[(X - E(X))\,{}^t(Y - E(Y))] = E(X\,{}^tY) - E(X)\,{}^tE(Y).$$

is called the *covariance* between X and Y. Then we have

$$\Lambda_Z = \begin{pmatrix} \Lambda_X & \Sigma_{XY} \\ {}^t\Sigma_{XY} & \Lambda_Y \end{pmatrix}.$$

Theorem 1 *If X and Y are independent normal vectors, then Z is normal and*

$$\Sigma_{XY} = 0.$$

Conversely, if Z is normal and if $\Sigma_{XY} = 0$, then X and Y are independent and normal.

Proof Let $t = \begin{pmatrix} u \\ v \end{pmatrix}$ be a vector of $\mathbb{R}^{p+q}$. We have

$$\langle t | Z \rangle = \langle u | X \rangle + \langle v | Y \rangle$$

and if X and Y are independent, we have

$$\varphi_Z(t) = \varphi_X(u)\varphi_Y(v), \tag{1}$$

i.e.,

$$\varphi_Z(t) = \exp\{i\langle u | m_X\rangle + i\langle v | m_Y\rangle - \tfrac{1}{2}\,{}^tu\Lambda_X u - \tfrac{1}{2}\,{}^tv\Lambda_Y v\} \tag{2}$$

and thus Z is a normal vector. Conversely, if Z is normal with $\Sigma_{XY} = 0$, then $\varphi_Z(t)$ is defined by (2) and (1) is satisfied so that X and Y are independent.

By induction, we can easily generalize this theorem to:

Theorem 2 *Let $X_1, \ldots, X_k$ be k random vectors and let ${}^tX$ be the vector $({}^tX_1, \ldots, {}^tX_k)$. Then the vector X is normal and the vectors X_i are uncorrelated, (i.e., $\Sigma_{X_iX_j} = 0$, $\forall i \neq j = 1, \ldots, k$) if and only if the vectors $X_1, \ldots, X_k$ are independent and normal.*

Theorem 3 *Suppose that Λ_X the covariance matrix of X is nonsingular; then the vector Z is normal if and only if the following conditions are satisfied:*

(a) *X is normal;*

(b) *the conditional distribution of Y given X is the normal distribution whose mean is*

$$E^x_{Y|X}(Y) = E(Y) + \Sigma_{YX}\Lambda_X^{-1}(x - E(X)) \tag{3}$$

and whose covariance matrix is

$$\Lambda_{Y|X}^{x} = \Lambda_Y - \Sigma_{YX}\Lambda_X^{-1}\Sigma_{XY}. \tag{4}$$

Note that the conditional expectation of Y given X is a linear function of X, but that the conditional covariance matrix of Y given X does not depend on X.

Proof (*Necessity*) We write

$$C = \Sigma_{YX}\Lambda_X^{-1}, \qquad Y' = Y - CX, \qquad Z' = \begin{pmatrix} X \\ Y' \end{pmatrix}.$$

Since Z is normal and Z' is a linear function of Z, then Z' is normal. On the other hand, X and Y' are uncorrelated since

$$E\{Y'\,{}^t X\} = E\{(Y - CX)\,{}^t X)\} = \Sigma_{YX} + E(Y')\,{}^t E(X) - C\Lambda_X.$$

Then using Theorem 1, we see that X and Y' are independent and the conditional distribution of Y' given X is the distribution of Y'. Since $Y = Y' + CX$, the conditional expectation of Y given X is

$$E_{Y|X}^{x}(Y) = E(Y') + Cx = E(Y) - CE(X) + Cx,$$

and the conditional covariance matrix of Y given X is

$$\Lambda_{Y|X}^{x} = \Lambda_{Y'}^{x} = \Lambda_{Y'}.$$

Proof (*Sufficiency*) There exist matrices Λ, A, b such that

$$E_{Y|X}^{x}(Y) = Ax + b, \qquad \Lambda_{Y|X}^{x} = \Lambda.$$

Let $Y'' = Y - AX - b$. Then

$$E_{Y''|X}^{x}(Y'') = 0, \qquad \Lambda_{Y''|X}^{x} = \Lambda.$$

The conditional distribution of Y'' given X is a normal distribution but does not depend on X. That is, the vectors X and Y'' are normal and independent. Using Theorem 1, we see that the vector Z'' is normal, where ${}^t Z'' = ({}^t X, {}^t Y'')$ and finally that the vector Z is also normal, since it is a linear function of Z''. ∎

Remark 1 In applications, to compute the conditional distribution of Y given X, one first determines the distribution of Z and then applies formulas (3) and (4).

Remark 2 The proof of the theorem shows that we can represent the conditional distribution of Y given X by

$$Y = E(Y) + \Sigma_{YX}\Lambda_X^{-1}(X - E(X)) + U,$$

where U is a normal vector which is independent of X and has zero expectation and a covariance matrix which is given by (4). This representation of the distribution of Y when X is given is called a *regression* function and we note that here it is a linear function. For a general case see the remark following Corollary A.4.4.

Remark 3 Let Y_i and Y_j be two components of Y. The correlation coefficient between Y_i and Y_j for the conditional distribution of Y given X is called the partial (or conditional) correlation between Y_i and Y_j given X. Using Formula (4) one can compute this coefficient easily. It does not depend on X.

Remark 4 When Y is a random variable, the coefficient $\Lambda^{x}_{Y|X}/\Lambda_Y$ is called the multiple correlation coefficient and generalizes the concept of correlation coefficient between two random variables.

Exercises

1. Let U be a random variable which can take only r distinct values, and let $X_1, \ldots, X_r$ be the number of occurrences of each value respectively among n independent trials of U.

(a) Show that we can consider the random vector X as the sum of n independent random vectors, where ${}^tX = ({}^tX_1, \ldots, {}^tX_r)$. Use this result to compute the generating function of X, and then, deduce from this its distribution.

(b) Compute the expectation and the covariance matrix of X.

(c) Find the distributions of X_1, of (X_1, X_2), of $X_1 + X_2$ and also the conditional distribution of (X_1, X_2) given $X_1 + X_2$.

(d) Compute the probability that $X_1, \ldots, X_m$ $(m < r)$ are all zeros. In the special case where U is uniformly distributed on its possible values, compute the probability of the event $\{X_1 > 0, \ldots, X_s > 0\}$ $(s \leqslant n)$.

(e) When n increases to infinity, find the limit of the distribution of $(1/n^{1/2})(X - E(X))$; describe it.

2. Compute the characteristic function, the variance and the moments of the $\Gamma(a, \lambda)$ distribution. Prove Theorem 1.1 and compute, when possible, the moments of the variables V and W.

Let $X_1, \ldots, X_n$ be n independent random variables which have respectively the $\Gamma(a_i, \lambda)$ $(i = 1, \ldots, n)$ distributions. Show that the random variables

$$Y_1 = \frac{X_1}{X_1 + X_2}, \quad Y_2 = \frac{X_1 + X_2}{X_1 + X_2 + X_3}, \ldots,$$

$$Y_{n-1} = \frac{X_1 + X_2 + \cdots + X_{n-1}}{X_1 + X_2 + \cdots + X_n}, \quad Y_n = X_1 + X_2 + \cdots + X_n$$

are independent and find the distribution of each.

3. Compute the characteristic function, the probability density function, the expectation and the variance of the $\Gamma(a, \gamma, \lambda)$ distribution. Show that the kth moment is a homogeneous polynomial of degree k with respect to γ and λ; compute its coefficients.

Let X be a random variable having a Poisson distribution and let Y be a random variable, such that the conditional distribution of Y given X is the $\Gamma(a + X, \lambda)$ distribution. Find the distribution of Y. Comment.

4. Let X be a normal random variable. Find the distribution of X^2. Let $X_1, \ldots, X_n$ be n independent random variables which have the same distribution $N(0, \sigma^2)$; find the distribution of $Y = X_1^2 + \cdots + X_n^2$. In the case where X is independent of Y and $E(X) = 0$ and $\sigma^2(X) = \sigma^2$ compute the probability density function (Student distribution) of the random variable $T = n^{1/2}(X/Y^{1/2})$. State and show corresponding results when $E(X) = m$.

5. Prove Theorem 1.2 and write the corresponding density of Y. Prove Theorem 1.3 and compute the density of T.

6. Let X be a random variable having the $\beta(a, b)$ distribution. Compute, when possible, the moments of X. Let Y be the random variable $(1 + X)^{-1}$; compute the probability density function of Y and the function $t \in \mathbb{R} \rightarrow \psi_Y(t) = E(Y^t)$. Let $Y_1, \ldots, Y_n$ be n independent random variables having the $\beta'(a_0 + a_1 + \cdots + a_{i-1}, a_i)$ $(i = 1, \ldots, n)$ distributions where $a_0, \ldots, a_n$ are given positive numbers. Then using the previous result, show that the product $Y_1 \cdots Y_n$ has a $\beta'(a_0, a_1 + \cdots + a_n)$ distribution. Give another proof of this result by using Exercise 2.

7. Let $X_1, X_2, \ldots, X_n$ be n independent normal random variables having zero expectations and unit variances. Find the distribution of the following sequence of random variables

$$Y_j = \frac{1}{(j(1 + j))^{1/2}} \sum_{i=1}^{j+1} (X_i - X_{j+1}), \qquad j = 1 \cdots n - 1,$$

$$Y_n = \frac{1}{n^{1/2}} \sum_{1}^{n} X_i,$$

and comment on the result.

8. Let X be a $(n + 1)$-dimensional normal vector. Designate its components by $Z, X_1, \ldots, X_n$. Assume that

$$E(Z) = E(X_1) = \cdots = E(X_n) = 0,$$

$$\sigma^2(Z) = \sigma^2(X_1) = \cdots = \sigma^2(X_n) = 1,$$

$$\rho(Z, X_i) = \alpha \qquad \text{for} \quad 1 \leqslant i \leqslant n,$$

$$\rho(X_i, X_j) = \alpha^2 \qquad \text{for} \quad 1 \leqslant i < j \leqslant n,$$

where α is a given number. Compute the distribution of X and deduce from it the distribution of the random variables $S = (1/n)\sum_1^n X_i$ and S/Z.

9. Let X be an n-dimensional normal vector and let A and B be two matrices of order (p, n) and (q, n) respectively. We designate by Λ the covariance matrix of X. Show that the random vectors AX and BX are independent if and only if

$$A\Lambda\,{}^t B = B\Lambda\,{}^t A = 0.$$

10. Prove Corollary 3.1.

11. Let $Z = \begin{pmatrix} X \\ Y \end{pmatrix}$ be a $(p + q)$-dimensional normal vector. Assume that the covariance matrix of Z is nonsingular.

(a) Compute $E({}^t XMY)$, where M is a given matrix of order (p, q).

(b) Show that there exists only one matrix A such that $E[(X - AY)^2]$ is minimized. What can you conclude from this?

12. Let X be a normal vector whose covariance matrix is the unit matrix.

(a) Find the distribution of $\|X\|^2$ and deduce from it the distribution of $\|X\|$. Prove this last result by a direct computation.

(b) Let S be the hypersphere whose center is $E(X)$ and whose radius is R; find R such that the uniform distribution *in* S has the same covariance matrix as X.

13. Let X be a normal vector having zero expectation, where ${}^t X = (X_1, \ldots, X_n)$. Find a method for computing $E(X_1^{a_1} \cdots X_n^{a_n})$, where $a_1, \ldots, a_n$ are given integers, and use this method to compute $E(X_1 X_2 \cdots X_n)$.

14. Let $Y_1, Y_2, \ldots, Y_n$ be n independent normal variables having zero expectations.

(a) Show that the equations $\begin{cases} Y_1 = X_1 \\ Y_2 = X_2 - a_1 X_1 \\ \vdots \\ Y_n = X_n - a_{n-1} X_{n-1} \end{cases}$

where $a_1, \ldots, a_{n-1}$ are given numbers, define a random vector ${}^t X = (X_1, \ldots, X_n)$. Compute the probability density function of X.

(b) Show that for every $k = 1, \ldots, n - 1$ the two random vectors

$${}^t U_k = (X_1, X_2, \ldots, X_k), \qquad {}^t V_k = (Y_{k+1}, \ldots, Y_n)$$

are independent.

(c) Let r_{ij} be the correlation coefficient between X_i and X_j; for every

$k = 1, \ldots, a-1$ compute the constant a_k by means of $r_{k,k+1}, r_{k-1,k}, \sigma^2(Y_k)$ and $\sigma^2(Y_{k+1})$.

(d) Show that for $j < i$

$$r_{ij} = r_{i,i-1} r_{i-1,i-2} \cdots r_{j+1,j}.$$

15. Let X be a normal vector, where ${}^tX = (X_1, X_2, \ldots, X_n)$, having the covariance matrix $\sigma^2 \mathbb{1}_n$.

(a) If $E(X) = 0$, find the conditional distribution of X given $\|X\|^2$.

(b) If $E({}^tX) = (m, \ldots, m)$, where m is a given number, find the conditional distribution of X given $(\bar{X}, \|X\|^2)$, where $\bar{X} = (1/n)\sum_1^n X_i$.

16. (a) Let $F(x, m, n)$ be the cumulative distribution function of the $\beta(m/2, n/2)$ distribution where m and n are given integers. Compute F by means of the cumulative distribution functions of binomial distributions. Then deduce from this an approximation of F by using the normal approximation to the binomial distributions.

(b) Do (a) when F is the cumulative distribution function of the $\beta'(m/2, n/2)$ distribution.

(c) Show that if n is a given integer, the cumulative distribution function $H(x, n, \lambda)$ of the $\Gamma(n, \lambda)$ distribution can be computed by means of the cumulative distribution functions of the Poisson distributions. Deduce from this an approximation formula.

(d) Deduce from the previous results an approximate relationship between F and H.

17. Let X be an n-dimensional normal vector and A and B two square symmetric matrices of order n. We designate by Λ the covariance matrix of X. Show that if $A\Lambda B = 0$, the random variables ${}^tXAX$ and ${}^tXBX$ are independent. Show that this result is also true if $E(X) = 0$ and $A\Lambda B\Lambda = 0$.

18. Let X and Y be two independent random vectors having the same distribution, $N(0, \Lambda)$.

(a) Show that $X + Y$ and $X - Y$ are independent and deduce from this the characteristic function of $\langle X | Y \rangle$. Verify this result by other means.

(b) Find the characteristic function of $(\langle X | Y \rangle, \|X\|^2)$.

19. Show that the following statistical space is complete

$$[\mathbb{R}^+, \mathscr{B}_{\mathbb{R}^+}; \Gamma(a, \alpha\gamma^2, \gamma), \gamma \in \mathbb{R}].$$

CHAPTER 8

Random Matrices

This chapter introduces matrices fundamental to multivariate analysis. To avoid complicated computations we introduce special, convenient and systematic notation. The use of characteristic functions will allow generalizations of certain definitions and simplification of proofs.

Having thus chosen adequate notations and methods, it will become obvious that the essential stochastic computations of multivariate analysis can be performed almost as easily as in the univariate case, provided that the techniques of matrix computation are used.

The format of this chapter corresponds to the standard treatment of the elementary univariate case and the step-by-step generalization of the elementary notions of covariance, of normal distributions, and of gamma distributions. The reader should note the central role played by the covariance, and its representation as the direct product of two symmetric matrices (Section 2). Most of the elementary properties generalize in a natural way.

1. Notation

We write $A[m, n]$ to designate a matrix A of order (m, n), i.e., a matrix which has m rows and n columns. However, after introducing this matrix, we shall often omit $[m, n]$. Furthermore, we always mean by $A = \{a_{ij}\}$ that a_{ij} is the term of A in the ith row and the jth column. When necessary, we shall associate, with every matrix $A[m, n]$, the vector $\tilde{A}$ belonging to $\mathbb{R}^{mn}$ defined by ${}^t\tilde{A} = ({}^tA_1, \ldots, {}^tA_n)$, where A_j is the jth column of A. The reader should note that $\tilde{A}$ is indeed a vector and not a partitioned matrix.

The inner product $\langle A | B \rangle$ of two matrices A and B, which have the same order, is defined as the Euclidean inner product of $\tilde{A}$ and $\tilde{B}$. We have

$$\langle A | B \rangle = \mathrm{Tr}(A\,{}^tB) = \mathrm{Tr}({}^tAB) = \mathrm{Tr}(B\,{}^tA) = \mathrm{Tr}({}^tBA) = \sum_{ij} a_{ij} b_{ij},$$

where the trace of a square matrix is the sum of its diagonal elements.

When convenient, with every symmetric matrix $A[r, r] = \{a_{ij}\}$ we associate the vector $\hat{A}$ belonging to $\mathbb{R}^{r(r+1)/2}$, defined by ${}^t\hat{A} = ({}^tT_1, \ldots, {}^tT_r)$, where the vector T_j belongs to $\mathbb{R}^j$ and is defined by

$${}^tT_j = (2^{1/2} a_{1j}, \ldots, 2^{1/2} a_{j-1,j}, a_{jj}) \qquad (j = 1, \ldots, r).$$

Note that if A and B are two symmetric matrices of the same order, the inner product of A and B is equal to the Euclidean inner product of $\hat{A}$ and $\hat{B}$ and both vector notations are consistent.

Let $A[m, n] = \{a_{ij}\}$ and $B[m', n'] = \{b_{hk}\}$ be two given matrices. We call the Kronecker or direct product of A and B, the matrix of order $[mm', nn']$, denoted by $A \otimes B$, where

$$A \otimes B = \begin{pmatrix} a_{11}B & \cdots & a_{1n}B \\ \vdots & & \vdots \\ a_{m1}B & \cdots & a_{mn}B \end{pmatrix}.$$

This definition means that the element of the matrix $A \otimes B$ which is in the $[(i-1)m' + h]$th row and in the $[(j-1)n' + k]$th column is equal to $a_{ij}b_{hk}$. Note that the Kronecker product is not commutative. We now give three of its properties:

T_1: Let A, B, C, D be four matrices of consistent orders; then

$$(A \otimes B)(C \otimes D) = (AC) \otimes (BD). \tag{1}$$

T_2: Let A and B be two square matrices; then

$$\mathrm{Tr}(A \otimes B) = \mathrm{Tr}(A)\,\mathrm{Tr}(B). \tag{2}$$

T_3: Let A and B be two nonsingular matrices; then

$$(A \otimes B)^{-1} = A^{-1} \otimes B^{-1}, \qquad \det(A \otimes B) = (\det A)^n (\det B)^m, \tag{3}$$

where m is the order of A and n the order of B.

The Lebesgue measure on the space $\mathcal{M}_{p,q}$ of all matrices of order $[p, q]$, is defined by

$$x = \{x_{ij}\} \in \mathcal{M}_{p,q}, \qquad dx = \prod_{i,j} dx_{ij},$$

and the Lebesgue measure on $\mathscr{S}_r$ is defined by

$$x = \{x_{ij}\} \in \mathscr{S}_r, \qquad dx = \prod_{i \leqslant j} dx_{ij}.$$

We denote by H_r the Haar invariant distribution on the space $\mathscr{C}_r$ of all orthogonal matrices of order r. This is (see Halmos [27, p. 262]) the unique distribution with the following property. If X is a random matrix having the distribution H_r, then for every orthogonal matrix C of the same order as X, the matrices XC and CX also have the distribution H_r. Let $\mathscr{C}_r^+$ be the subset of $\mathscr{C}_r$ in which all elements of the first column are nonnegative. We define the conditional Haar invariant distribution on $\mathscr{C}_r^+$ as the distribution equal to 2^r times the Haar invariant distribution on this subset.

For every integer r and every real number a, such that $r \geqslant 1$, $2a + 1 > r$, we write

$$\Gamma_r(a) = \pi^{r(r-1)/4} \prod_{j=0}^{r-1} \Gamma(a - \tfrac{1}{2}j).$$

Note that for $r = 1$, we obtain the usual gamma function. Finally, for every square matrix x of order r, we denote by $R(x)$, the sequence of eigenvalues of x in decreasing order,

$$R(x) = \{R_1(x) \geqslant \cdots \geqslant R_r(x)\}. \tag{4}$$

2. Covariance and Characteristic Function of a Random Matrix

Definition 1 Let $U[p, q]$ be a random matrix. We call Φ_U, the following function on $\mathscr{M}_{p,q}$, the *characteristic function* of U:

$$T \in \mathscr{M}_{p,q} \to \Phi_U(T) = E(\exp i\langle T \mid U\rangle).$$

This is consistent with the definition of the characteristic function of $\tilde{U}$. When the matrix U belongs to $\mathscr{S}_r$, we can restrict the definition of Φ_U to $\mathscr{S}_r$ and the definition is then consistent with that of the characteristic function of $\hat{U}$.

Theorem 1 *Let X be a random matrix and let A and B be two given matrices which have orders enabling the matrix $Y = AXB$ to be defined. Then we have*

$$\Phi_Y(T) = \Phi_X({}^t A T\, {}^t B).$$

Proof The proof of this theorem follows from

$$\mathrm{Tr}(T\, {}^t(AXB)) = \mathrm{Tr}(T\, {}^t B\, {}^t X\, {}^t A) = \mathrm{Tr}({}^t A T\, {}^t B\, {}^t X).$$

Specifically if X is a symmetric random matrix, the random matrix $Y = AX\,{}^{t}A$ is also symmetric, and we have $\Phi_Y(T) = \Phi_X({}^{t}ATA)$.

Definition 2 Let $U[p, q]$ be a random matrix such that the square of each element of U is integrable; then we call the *covariance* of U, the quadratic form Λ_U defined on $\mathscr{M}_{p,q}$, where

$$\forall x \in \mathscr{M}_{p,q}, \qquad \Lambda_U(x) = E(\langle U - E(U) | x \rangle^2) = {}^{t}\tilde{x}\Lambda_{\tilde{U}}\tilde{x}.$$

If U is a symmetric matrix, we can restrict the definition of Λ_U to $\mathscr{S}_r$.

Theorem 2 *If X has a covariance, then Y, as defined in Theorem* 1, *also has a covariance, and*

$$\Lambda_Y(y) = \Lambda_X({}^{t}Ay\,{}^{t}B).$$

Definition 3 Let $U[p, q] = \{U_{ij}\}$ be a random matrix which has zero expectation and covariance Λ_U. We say that the covariance Λ_U is a *direct product* if there exist two matrices $\Lambda' = \{\lambda'_{ii'}\}$ and $\Lambda'' = \{\lambda''_{jj'}\}$ belonging to $\mathscr{S}_p^+$ and $\mathscr{S}_q^+$, respectively, such that one of the following three equivalent conditions is fulfilled

$$\Lambda_{\tilde{U}} = \Lambda'' \otimes \Lambda' \tag{1}$$

$$E(U_{ij}U_{i'j'}) = \lambda'_{ii'}\lambda''_{jj'} \tag{2}$$

$$\forall z \in \mathscr{M}_{p,q}, \qquad \Lambda_U(z) = \langle z\Lambda'' | \Lambda' z \rangle = \mathrm{Tr}(z\Lambda''\,{}^{t}z\Lambda'). \tag{3}$$

The proof of the equivalences is easy and left as an exercise for the reader. When the random matrix U satisfies these conditions, we say that Λ_U is the direct product of Λ' and Λ''.

Using Theorem 2 and Condition (3) we can prove

Theorem 3 *Let U be a random matrix whose covariance is the direct product of Λ' and Λ''. Then the random matrix $Y = AUB$, where A and B are given matrices with convenient orders, has a covariance which is the direct product of $A\Lambda'\,{}^{t}A$ and of ${}^{t}B\Lambda''B$.*

Corollary *Let $X[p, q]$ be a random matrix whose covariance is the direct product of Λ' and Λ''. Then there exist two matrices $A[p, \mathrm{rank}(\Lambda')]$, $B[q, \mathrm{rank}(\Lambda'')]$ such that $X = AUB$, where U is a random matrix whose elements are uncorrelated.*

Proof We use Theorem 3 with Λ' and Λ'' equal to unit matrices, where matrices A and B are obtained by application of Lemma 7.2.1. ∎

3. Some Miscellaneous Results

Let A be an open set of $\mathscr{M}_{p,q}$ and f a one-to-one differentiable mapping from A into $\mathscr{M}_{p,q}$. We define the Jacobian of the transformation between x and $y = f(x)$ as the Jacobian of the corresponding transformation between $\tilde{x}$ and $\tilde{y}$; i.e.,

$$J = \left|\frac{\partial y}{\partial x}\right| = \left|\det\left(\frac{\partial \tilde{y}}{\partial \tilde{x}}\right)\right|,$$

where

$$\tilde{y} = g(\tilde{x}) \quad \Leftrightarrow \quad y = f(x).$$

If x and y are symmetric or upper triangular matrices we use variables $\{x_{ij}, i \leqslant j\}$ and $\{y_{ij}, i \leqslant j\}$ instead of $\tilde{x}$ and $\tilde{y}$; in such cases the reader will note that for consistency with the definition of Lebesgue measure on $\mathscr{S}_r$ we do not use $\hat{x}$ and $\hat{y}$.

We now give Jacobians of some useful transformations.

Theorem 1 *Let A be a nonsingular matrix of order n; the transformation on $\mathscr{M}_{n,m}$ defined by $y = Ax$ has the Jacobian $|\partial y/\partial x| = |\det(A)|^m$.*

Proof It is easy to see that the mapping $\tilde{x} \to \widetilde{Ax}$ is linear and defined by the matrix $\mathbb{1}_m \otimes A$. The proof is completed by using (1.3). ∎

Note that an analogous result holds for the transformation $y = xA$.

Theorem 2 *Let A be a nonsingular matrix of order n; the transformation on $\mathscr{S}_n$ defined by $y = Ax\,{}^tA$ has the Jacobian $|\partial y/\partial x| = |\det A|^{n+1}$.*

Theorem 3 *Let $\mathscr{I}_n$ be the space of nonsingular matrices of order n; then the transformation on $\mathscr{I}_n$ defined by $y = x^{-1}$ has the Jacobian $|\partial y/\partial x| = |\det x|^{-2n}$.*

Proof The derivative of $x \to x^{-1}$ at the point x_0 [13; p. 148] is the mapping,

$$u \to -x_0^{-1} u x_0^{-1}.$$

The proof follows by using Theorem 1 on the left and right sides. ∎

Theorem 2 can be used to demonstrate

Theorem 4 *Let Σ_n be the space of symmetric nonsingular matrices of order n; the transformation on Σ_n defined by $y = x^{-1}$ has the Jacobian $|\partial y/\partial x| = |\det x|^{-n-1}$.*

Theorem 5 *Let x be a matrix in $\mathscr{S}_n^+$. There exists a unique triangular matrix $T = \{t_{ij}\}$ such that $x = T\,{}^{t}T$ and the diagonal elements of T are nonnegative. The mapping $x \to T$ has the Jacobian*

$$\left|\frac{\partial x}{\partial T}\right| = 2^n t_{11} t_{22}^2 \cdots t_{nn}^n.$$

In this theorem, triangular means that all elements of T below the diagonal are zero. If, on the other hand, we assume that triangular means that all elements above the diagonal are zero, the Jacobian becomes

$$2^n t_{11}^n t_{22}^{n-1} \cdots t_{nn}.$$

Theorem 6 (Anderson, [1; p. 313]) *Let A and B be two matrices belonging to $\mathscr{S}_r^+$, such that the eigenvalues of AB^{-1} are different. Then there exist (a.s.) a unique nonsingular matrix E of order r whose first column is positive, and a diagonal matrix F whose diagonal elements f_j $(j = 1, \ldots, r)$ are in decreasing order, such that*

$$A = {}^{t}EFE, \qquad B = {}^{t}E(1 - F)E. \tag{1}$$

The Jacobian of the transformation defined by (1) *is*

$$\left|\frac{\partial(A, B)}{\partial(E, F)}\right| = 2^r |\det E|^{r+2} \prod_{1 \leq i < j \leq r} (f_i - f_j).$$

This theorem is the generalization to the matrix case of the elementary transformation

$$(x > 0, y > 0) \to (x + y, x/y)$$

Theorem 7 [48; p. 54] *Let A and B be two square matrices of the same order; then AB and BA have the same characteristic roots.*

Corollary *Let $A[m, n]$ and $B[n, m]$ be two given matrices; then we have*

$$\forall s \in \mathbb{R}, \qquad \det(\mathbb{1}_m - sAB) = \det(\mathbb{1}_n - sBA).$$

Proof Add zero elements to A and B such that they become square matrices and apply Theorem 7. ∎

4. Fundamental Results

Some of the following proofs refer to later results. However, the reader should observe, that the proofs remain consistent. We shall need the following lemma, whose proof is left to the reader.

Lemma *Let X and Y be two random elements taking values on $(\mathscr{X}, \mathscr{C})$ and let f be a measurable mapping from $(\mathscr{X}, \mathscr{C})$ to $(\mathscr{Y}, \mathscr{D})$. If P_X is absolutely continuous with respect to P_Y, then $P_{f(X)}$ is absolutely continuous with respect to $P_{f(Y)}$, and we have*

$$\frac{dP_{f(X)}}{dP_{f(Y)}}(f(Y)) = E\left[\frac{dP_X}{dP_Y}(Y) \mid f(Y)\right]. \tag{1}$$

Specifically if

$$\frac{dP_X}{dP_Y} = g(f),$$

then

$$\frac{dP_{f(X)}}{dP_{f(Y)}} = g.$$

Theorem 1 *Let $X[r, m]$ $(r \leqslant m)$ be a random matrix whose probability density function (with respect to the Lebesgue measure on $\mathscr{M}_{r,m}$) depends only on $x\,{}^{t}x$ and is designated as $h(x\,{}^{t}x)$. Then the random matrix $Z = X\,{}^{t}X$ has the probability density function*

$$\frac{\pi^{rm/2}}{\Gamma_r(m/2)}(\det z)^{\frac{1}{2}(m-r-1)}h(z).$$

Proof Apply the previous Lemma with

$$P_Y = N(O, \mathbb{1}_r \otimes \mathbb{1}_m) \qquad \text{(see Section 5)}$$

and

$$f : x \to x\,{}^{t}x.$$

Using Theorem 5.4, we have

$$\frac{dP_Y}{dy} = (2\pi)^{-rm/2} \exp[-\tfrac{1}{2}\operatorname{Tr}(y\,{}^{t}y)].$$

Then, Theorem 6.1 and Theorem 8.1 show that

$$\frac{dP_{Y\,{}^{t}Y}}{dz} = \frac{2^{-rm/2}}{\Gamma_r(m/2)} \exp(-\tfrac{1}{2}\operatorname{Tr} z)(\det z)^{(m-r-1)/2}.$$

Thus,

$$\frac{dP_X}{dP_Y} = h(x\,{}^{t}x)(2\pi)^{rm/2} \exp[\tfrac{1}{2}\operatorname{Tr}(x\,{}^{t}x)].$$

Here the function g of the lemma is equal to

$$(2\pi)^{rm/2}\exp(\tfrac{1}{2}\operatorname{Tr} z)h(z),$$

and

$$\frac{dP_Z}{dP_{Y\,{}^tY}} = (2\pi)^{rm/2}\exp(\tfrac{1}{2}\operatorname{Tr} z)h(z).$$

Finally,

$$\frac{dP_Z}{dz} = \left(\frac{dP_{Y\,{}^tY}}{dz}\right)\left(\frac{dP_Z}{dP_{Y\,{}^tY}}\right)$$ ∎

Theorem 2 *Let X be a random matrix belonging to $\mathscr{S}_r$ whose probability density function (with respect to the Lebesgue measure on $\mathscr{S}_r$) depends only on $R(x)$ (as defined by* (1.4), *and is designated by $f(R)$. Then,*

(1) *The probability density function of $R(X)$ is equal to*

$$\frac{\pi^{r^2/2}}{\Gamma_r(r/2)} f(R)\prod_{i<j}(R_i - R_j) \qquad (R_1 \geqslant \cdots \geqslant R_r).$$

(2) *The distribution of the unique orthogonal matrix $C(X)$ whose first column is positive and satisfies*

$$X = {}^tC(X)D(X)C(X),$$

where $D(X)$ is the diagonal matrix consisting of the elements of $R(X)$, is the conditional Haar invariant distribution. Moreover $R(X)$ and $C(X)$ are independent.

Proof (1) Apply the previous lemma with

$$f : x \to R(x), \qquad P_Y = B_r(a, r/2).$$

Using Theorem 9.1, we have

$$\frac{dP_Y}{dy} = \frac{\Gamma_r(a + \frac{1}{2}r)}{\Gamma_r(a)\Gamma_r(r/2)}(\det x)^{a-(r+1)/2}\det(\mathbb{1}_r + x)^{-a-r/2}.$$

Finally use Theorem 9.7 to obtain the probability density function of $R(Y)$,

$$\frac{\pi^{r^2/2}\Gamma_r(a+\frac{1}{2}r)}{\Gamma_r(a)\Gamma_r^2(r/2)}\left(\prod_1^r u_i\right)^{a-(r+1)/2}\left(\prod_1^r(1+u_i)\right)^{-a-r/2}\prod_{1\leqslant i<j<r}(u_i - u_j).$$

(2) For every orthogonal matrix H, the random matrix ${}^tHXH$ has the same distribution as X. Then, for every orthogonal matrix H, the conditional distribution of $C(X)$ given $R(X)$ is invariant under the transformation

which associates to C the matrix obtained from CH by changing the sign of each row, if necessary, to make its first column nonnegative. This property characterizes the conditional Haar invariant distribution (see Exercise 4). Then the conditional distribution of $C(X)$ given $R(X)$ does not depend on $R(X)$. ∎

5. Normal Random Matrix

We designate by $N(M, \Lambda \otimes \Lambda')$ the distribution of a random matrix X, when the associated vector $\tilde{X}$ has the distribution

$$N(\tilde{M}, \Lambda' \otimes \Lambda).$$

Theorem 2.3 can be used to show:

Theorem 1 *Let X be a random matrix whose distribution is $N(M, \Lambda \otimes \Lambda')$ and let A and B be two given matrices which have orders permitting the matrix $Y = AXB$ to be defined. Then the distribution of Y is*

$$N(AMB, (A\Lambda\,{}^{t}A) \otimes ({}^{t}B\Lambda' B)).$$

Using this theorem and Lemma 7.2.1, one can show

Theorem 2 *Let $X[m, n]$ be a random matrix having the $N(M, \Lambda \otimes \Lambda')$ distribution and let $A[m, \operatorname{rank}(\Lambda)]$ and $B[\operatorname{rank}(\Lambda'), n]$ be two matrices such that*

$$\Lambda = A\,{}^{t}A, \qquad \Lambda' = {}^{t}BB.$$

Then we can write $X = M + AUB$, where the elements of the random matrix U are independent, standard normal random variables.

It is easy to see that if the random matrix $M[p, q] = \{M_{ij}\}$ has the $N(O, \mathbb{1}_p \otimes \mathbb{1}_q)$ distribution, the random variables M_{ij} are independent, normal and standardized. If the distribution of M is $N(O, \Lambda \otimes \mathbb{1}_q)$, the columns of M are independent, each having the $N(O, \Lambda)$ distribution. If the distribution of M is $N(O, \mathbb{1}_p \otimes \Lambda')$, the rows of M are independent, each having the $N(O, \Lambda')$ distribution.

Using (2.3), one can show:

Theorem 3 *The characteristic function of the $N(M, \Lambda \otimes \Lambda')$ distribution is equal to*

$$T \to \exp\{i\langle T \mid M\rangle - \tfrac{1}{2}\langle T\Lambda' \mid \Lambda T\rangle\}.$$

Theorem 4 *If the two matrices $\Lambda \in \mathscr{S}_m^+$ and $\Lambda' \in \mathscr{S}_n^+$ are nonsingular, then the $N(O, \Lambda \otimes \Lambda')$ distribution has a probability density function (with respect to Lebesgue measure on $\mathscr{M}_{m,n}$), equal to*

$$x \to (2\pi)^{-mn/2}(\det \Lambda)^{-n/2}(\det \Lambda')^{-m/2} \exp\{-\tfrac{1}{2}\langle x\Lambda'^{-1} | \Lambda^{-1}x\rangle\}.$$

Proof The normal vector $\tilde{X}$ has the probability density function,

$$\tilde{x} \to (2\pi)^{-mn/2}[\det(\Lambda' \otimes \Lambda)]^{-1/2} \exp\{-\tfrac{1}{2}\,{}^t\tilde{x}(\Lambda' \otimes \Lambda)^{-1}\tilde{x}\}.$$

Using (1.3) we see that this function is equal to

$$\tilde{x} \to (2\pi)^{-mn/2}(\det \Lambda)^{-n/2}(\det \Lambda')^{-m/2} \exp\{-\tfrac{1}{2}\,{}^t\tilde{x}(\Lambda'^{-1} \otimes \Lambda^{-1})\tilde{x}\}.$$

But the equivalence of (2.1) and (2.3) shows that

$${}^t\tilde{x}(\Lambda'^{-1} \otimes \Lambda^{-1})\tilde{x} = \langle x\Lambda'^{-1} | \Lambda^{-1}x\rangle. \quad \blacksquare$$

6. Generalized Gamma Distributions

We now generalize the gamma distribution given in Definition 7.1. For statistical purposes we need study only the case where $2a$ is an integer. However when possible we have proved the properties for general a. On the other hand, the reader should keep in mind that the underlying notion here is that of a random quadratic form.

Definition 1 Let a and the integer r be two positive numbers and let Λ be a matrix belonging to $\mathscr{S}_r^+$. When it exists, we designate by $\Gamma_r(a, \Lambda)$ the distribution on $\mathscr{S}_r^+$ which has the characteristic function

$$[\det(\mathbb{1}_r - iT\Lambda)]^{-a}, \qquad T \in \mathscr{S}_r^+.$$

Note that, by using Theorem 3.7 one can write ΛT instead of $T\Lambda$. The $\Gamma_r(a, \Lambda)$ distribution is called a *generalized gamma distribution*. The following theorem shows that if $2a$ is an integer, then $\Gamma_r(a, \Lambda)$ exists and is called the Wishart distribution.

Theorem 1 *Let $X[n, r]$ be a random matrix having the $N(O, \mathbb{1}_n \otimes \Lambda)$ distribution; then the random matrix ${}^tXX$ has the $\Gamma_r(n/2, 2\Lambda)$ distribution.*

Proof The rows $X_1, \ldots, X_n$ of X are independent normal vectors having the same covariance Λ. Then we have

$$\Phi_{{}^tXX}(T) = E(\exp[i \operatorname{Tr}({}^tXXT)]) = [\Phi_{X_1T{}^tX_1}(1)]^n,$$

since

$$\operatorname{Tr}({}^t XXT) = \operatorname{Tr}\left(\left(\sum_1^n {}^t X_i X_i\right) T\right) = \operatorname{Tr}\left(\sum_1^n {}^t X_i X_i T\right) = \sum_1^n \operatorname{Tr}({}^t X_i X_i T)$$
$$= \sum_1^n \operatorname{Tr}(X_i T\,{}^t X_i) = \sum_1^n X_i T\,{}^t X_i.$$

The proof is completed by using Theorem 7.3.1. ∎

We remark that if $\Lambda = \mathbb{1}_r$, the two matrices ${}^t XX$ and $X\,{}^t X$ have generalized gamma distributions. It will be shown later that one of these distributions is singular.

Corollary (Wishart and Bartlett) *Let $X_1, \ldots, X_n$ be n independent normal vectors, each having the $N(m, \Lambda)$ distribution on $\mathbb{R}^r$. Then the random vector $\bar{X} = (1/n)\sum_1^n X_i$ and the random matrix $S = (1/n)\sum_1^n X_i\,{}^t X_i - \bar{X}\,{}^t\bar{X}$ are independent; $\bar{X}$ has the $N(m, \Lambda/n)$ distribution and S has the $\Gamma_r((n-1)/2, 2\Lambda/n)$ distribution.*

Proof Without loss of generality we can assume that $m = 0$. Let C be an orthogonal matrix which has all elements of the first column equal to $n^{-1/2}$; then the two random matrices

$$X = (X_1, \ldots, X_n), \qquad Y = XC = (Y_1, \ldots, Y_n)$$

have the same distribution, $N(O, \Lambda \otimes \mathbb{1}_n)$. Furthermore, we have $\bar{X} = n^{-1/2} Y_1$,

$$nS = X\,{}^t X - n\bar{X}\,{}^t\bar{X} = Y\,{}^t Y - Y_1\,{}^t Y_1 = \sum_2^n Y_j\,{}^t Y_j$$

and the proof is completed by using Theorem 1. ∎

Theorem 2 $\Gamma_r(a, \Lambda) * \Gamma_r(b, \Lambda) = \Gamma_r(a + b, \Lambda)$.

Theorem 3 *Let X be a random matrix having the $\Gamma_r(a, \Lambda)$ distribution and let $A[r, m]$ be a given matrix, then the random matrix $Y = {}^t AXA$ has the $\Gamma_m(a, {}^t A\Lambda A)$ distribution.*

Proof Using Theorem 2.1 we have

$$\Phi_Y(T) = \Phi_X(AT\,{}^t A).$$

Then

$$\Phi_Y(T) = [\det(\mathbb{1}_r - iAT\,{}^t A\Lambda)]^{-a}$$

and the proof is completed by using the corollary of Theorem 3.7. ∎

Theorem 4 *Let X be a random matrix having $\Gamma_r(a, \Lambda)$ distribution. If X and Λ are partitioned in blocks of the same order*

$$X = \begin{pmatrix} X' & X'' \\ {}^t X'' & X''' \end{pmatrix}, \qquad \Lambda = \begin{pmatrix} \Lambda' & \Lambda'' \\ {}^t\Lambda'' & \Lambda''' \end{pmatrix},$$

where X' and Λ' are of order r', then the distributions of X' and X''' are $\Gamma_{r'}(a, \Lambda')$ and $\Gamma_{r-r'}(a, \Lambda''')$ respectively. Furthermore, if $\Lambda'' = 0$, the matrices X' and X''' are independent.

Proof The first part of Theorem 4 is proved by using Theorem 3 twice, with matrices

$$A_1 = \begin{pmatrix} \mathbb{1}_{r'} \\ 0 \end{pmatrix}, \qquad A_2 = \begin{pmatrix} 0 \\ \mathbb{1}_{r-r'} \end{pmatrix}.$$

The characteristic function of the pair (X', X''') is equal to

$$T' \in \mathscr{S}_{r'}, \qquad T''' \in \mathscr{S}_{r-r'} \to \Phi_X\left[\begin{pmatrix} T' & 0 \\ 0 & T''' \end{pmatrix}\right].$$

Since $\Lambda'' = 0$ here, this characteristic function is equal to

$$\det\left[\mathbb{1}_r - i\begin{pmatrix} T' & 0 \\ 0 & T''' \end{pmatrix}\begin{pmatrix} \Lambda' & 0 \\ 0 & \Lambda''' \end{pmatrix}\right] = [\det(\mathbb{1}_{r'} - iT'\Lambda')][\det(\mathbb{1}_{r-r'} - iT''\Lambda''')]. \quad \blacksquare$$

7. Bartlett's Decomposition of a Gamma Distribution

Lemma 1 *Let X be a random vector belonging to $\mathbb{R}^{p-1}$, having the $N(O, \Lambda)$ distribution and let Y be a random variable which is independent of X. Assume that Y^2 has the $\Gamma(a, \lambda)$ distribution and let*

$$Z = \begin{pmatrix} X \\ Y \end{pmatrix}, \qquad U = Z\,{}^tZ, \qquad \Sigma = \begin{pmatrix} 2\Lambda & 0 \\ 0 & \lambda \end{pmatrix}.$$

Then the characteristic function of U is equal to

$$T \in \mathscr{S}_p \to [\det(\mathbb{1}_{p-1} - 2iT'\Lambda)]^{a-\frac{1}{2}}[\det(\mathbb{1}_p - iT\Sigma)]^{-a}$$

where T' is the matrix obtained from T by deleting the last column and the last row.

Proof We have

$$\Phi_U(T) = E(E(\exp[i\,{}^tZTZ]\,|\,Y)). \tag{1}$$

The conditional distribution of Z given Y is

$$N\left(\begin{pmatrix}0\\Y\end{pmatrix}, \begin{pmatrix}\Lambda & 0\\0 & 0\end{pmatrix}\right)$$

so that we can compute the conditional expectation

$$E(\exp[i\,{}^{t}ZTZ] \mid Y). \tag{2}$$

Using Theorem 7.3.1, with†

$$n = p, \quad t = 1, \quad M = T, \qquad m = \begin{pmatrix}0\\Y\end{pmatrix}, \qquad \Lambda' = \begin{pmatrix}\Lambda & 0\\0 & 0\end{pmatrix},$$

we obtain for (2)

$$\left[\det\left(\mathbb{1}_p - 2iT\begin{pmatrix}\Lambda & 0\\0 & 0\end{pmatrix}\right)\right]^{-1/2} \exp\left\{iY^2(0,1)\left[\mathbb{1}_p - 2iT\begin{pmatrix}\Lambda & 0\\0 & 0\end{pmatrix}\right]^{-1} T\begin{pmatrix}0\\1\end{pmatrix}\right\}. \tag{3}$$

The expectation of (3) requires calculation of only the characteristic function of Y^2 and we obtain for (1),

$$\left[\det\left(\mathbb{1}_p - 2iT\begin{pmatrix}\Lambda & 0\\0 & 0\end{pmatrix}\right)\right]^{-1/2}\left[1 - i\lambda(0,1)\left[\mathbb{1}_p - 2iT\begin{pmatrix}\Lambda & 0\\0 & 0\end{pmatrix}\right]^{-1} T\begin{pmatrix}0\\1\end{pmatrix}\right]^{-a}.$$

By an elementary computation we have

$$[\det(\mathbb{1}_{p-1} - 2iT'\Lambda)]^{-1/2}[1 - i\lambda\beta + 2\lambda\,{}^{t}\alpha\Lambda(\mathbb{1}_{p-1} - 2iT'\Lambda)^{-1}\alpha],$$

where

$$T = \begin{pmatrix}T' & \alpha\\ {}^{t}\alpha & \beta\end{pmatrix}.$$

Finally, the proof is completed by using the property,

$$\det\begin{pmatrix}M & U\\V & \gamma\end{pmatrix} = \det M \det(\gamma - VM^{-1}U), \tag{4}$$

where M is nonsingular and γ is a real number. ∎

Theorem 1 *Let p be a positive integer, a be a real number such that $2a + 1 > p$ and let $X = \{X_{ij}\}$ be a triangular matrix of order p whose ele-*

† Note that Λ' appeared as Λ in Theorem 7.3.1. This circumlocution is necessary because Λ is different in the two theorems. (Ed.)

ments are independent. If the random variable X_{ij} $(j > i)$ *has the* $N(O, \frac{1}{2})$ *distribution and if the random variable* X_{ii}^2 *has the* $\Gamma(a + \frac{1}{2}(i - p), 1)$ *distribution, then the random matrix* $X\,{}^tX$ *has the* $\Gamma_p(a, \mathbb{1}_p)$ *distribution.*

Proof Let ${}^tX_j (j = 1, \ldots, p)$ be the random vector $(X_{1j}, \ldots, X_{jj})$. Then we have

$$X\,{}^tX = \sum_{j=1}^{p} \begin{pmatrix} X_j\,{}^tX_j & 0 \\ 0 & 0 \end{pmatrix}.$$

The characteristic function of $X_j\,{}^tX_j$ is given by Lemma 1 and the matrices

$$\begin{pmatrix} X_j\,{}^tX_j & 0 \\ 0 & 0 \end{pmatrix} \qquad (j = 1, \ldots, p)$$

are independent. We can now compute the characteristic function of $X\,{}^tX$. ∎

Corollary 1 *If* $2a + 1 > \operatorname{rank} \Lambda$, *the distribution* $\Gamma_r(a, \Lambda)$ *exists.*

Proof Using the Lemma of Section 7.2, there exists a matrix $A[r, \operatorname{rank}(\Lambda)]$ such that $A\,{}^tA = \Lambda$. On the other hand, Theorem 1 shows that the $\Gamma_p(a, \mathbb{1}_p)$ distribution exists when $2a + 1 > p$. The proof is completed by using Theorem 6.3. ∎

Corollary 2 *If* Λ *is nonsingular, the distribution of the random variable* $\det X / \det \Lambda$, *where* X *has the* $\Gamma_r(a, \Lambda)$ *distribution, is the distribution of the product of* r *independent random variables having* $\Gamma(a - \frac{1}{2}s, 1)$ $(s = 0, \ldots, r - 1)$ *distributions.*

Proof The distribution of $\det X / \det \Lambda$ is that of $\det X'$ where X' has the $\Gamma_r(a, \mathbb{1}_r)$ distribution. The proof is completed by using Theorem 1. ∎

8. Nonsingular Gamma Distribution

Theorem 1 *Let* Λ *be a nonsingular matrix belonging to* $\mathscr{S}_r^+$. *If* $2a + 1 > r$, *the* $\Gamma_r(a, \Lambda)$ *distribution has a density (with respect to Lebesgue measure on* $\mathscr{S}_r^+$*)*

$$x \in \mathscr{S}_r^+ \to \frac{(\det \Lambda)^{-a}}{\Gamma_r(a)} \exp[-\langle x | \Lambda^{-1} \rangle](\det x)^{a-(r+1)/2}.$$

Proof First consider the special case where $\Lambda = \mathbb{1}_r$ and apply Theorem

7.1 (assuming that $X_{ii} > 0$). The joint probability density function of the variables X_{ij} is

$$\prod_{i<j} (\pi)^{-1/2} \exp[-(x_{ij})^2] \prod_j \frac{2\exp[-(x_{jj})^2]}{\Gamma(a+\frac{1}{2}(j-r))} (x_{jj})^{2a-r+j-1}$$
$$= \frac{2^r}{\Gamma_r(a)} \exp\left[-\sum_{i\leqslant j}(x_{ij})^2\right]\left(\prod_j x_{jj}\right)^{2a-r-1} \prod_j (x_{jj})^j.$$

Using Theorem 3.5 we have

$$2^r \prod_j (x_{jj})^j = \frac{dX\,{}^tX}{dX}$$

and then the probability density function of $X\,{}^tX$ is equal to

$$y \to \frac{1}{\Gamma_r(a)} \exp[-\operatorname{Tr} y](\det y)^{a-(r+1)/2}, \tag{1}$$

since

$$\sum_{i\leqslant j} (x_{ij})^2 = \operatorname{Tr}(X\,{}^tX), \qquad \det(X\,{}^tX) = (\det X)^2 = \prod_j x_{jj}.$$

Now consider the general case and let A be a nonsingular matrix such that

$$\Lambda = A\,{}^tA.$$

Let Y be a random matrix having the $\Gamma_r(a, \mathbb{1}_r)$ distribution; then the random matrix $Z = AY\,{}^tA$ has the $\Gamma_r(a, \Lambda)$ distribution. However, Theorem 3.2 shows that the Jacobian dZ/dY is equal to $[\det A]^{r+1}$, i.e., to $(\det \Lambda)^{(r+1)/2}$. From (1), compute the probability density of Z and the proof is completed since

$$\operatorname{Tr} Y = \operatorname{Tr}(A^{-1}X\,{}^tA^{-1}) = \operatorname{Tr}(X\,{}^tA^{-1}A^{-1}) = \operatorname{Tr}(X(A\,{}^tA)^{-1}) = \operatorname{Tr}(X\Lambda^{-1})$$
$$\det X = \det Y(\det A)^2 = \det Y \det \Lambda. \qquad \blacksquare$$

Theorem 2 *If $2a + 1 > r$, the statistical space*

$$[\mathscr{S}_r^+, \mathscr{B}_{\mathscr{S}_r^+}; \Gamma_r(a, \Lambda), \Lambda \in \mathscr{S}_r^+] \tag{2}$$

is complete.

Proof Let m be the measure on $\mathbb{R}^{r(r+1)/2}$ whose density (with respect to Lebesgue measure) is equal to $(\det x)^{a-(r+1)/2}$ on the cone of vectors $\tilde{x}$ corresponding to a matrix x belonging to $\mathscr{S}_r^+$. The measure m is a smooth measure (see Definition 10.1.3). Then the statistical space (2) is the exponential statistical space associated with m (see Definition 10.2.2). Using Theorem 10.2.1 completes the proof.

Theorem 3 *Let X be a random matrix having the $\Gamma_r(a, \mathbb{1}_r)$ distribution. If $2a + 1 > r$, then $\gamma_r(a)$, the distribution of $R(X)$ defined by (1.4), has the probability density function,*

$$\frac{\pi^{r^2/2}}{\Gamma_r(a)\Gamma_r(r/2)} \exp\left[-\sum_1^r u_i\right]\left(\prod_1^r u_i\right)^{a-(r+1)/2} \prod_{i<j} (u_i - u_j),$$

where $u_1 > \cdots > u_r > 0$.

This theorem is a straightforward application of Theorem 4.2.

9. Generalized Beta Distributions

Theorem 1 *Let a and b be two real numbers and r a positive integer such that*

$$2a + 1 > r, \qquad 2b + 1 > r.$$

Let Y be a random matrix having the $\Gamma_r(b, \mathbb{1}_r)$ distribution and let X be a random matrix such that the conditional distribution of X given Y is the $\Gamma_r(a, Y^{-1})$ distribution. Then the distribution of X, which is designated by $B_r(a, b)$, has the probability density function (with respect to Lebesgue measure on $\mathscr{S}_r^+$)

$$\frac{\Gamma_r(a+b)}{\Gamma_r(a)\Gamma_r(b)} (\det x)^{a-(r+1)/2} [\det(\mathbb{1}_r + x)]^{-a-b}.$$

Proof Since the random matrix Y is almost everywhere nonsingular, the conditional probability density function of X given Y is equal to

$$\frac{1}{\Gamma_r(a)} (\det y^{-1})^{-a} \exp[-\mathrm{Tr}(xy)](\det x)^{a-(r+1)/2}. \tag{1}$$

The probability density function of X is the expectation of (1) with respect to the distribution of Y. Thus this function is equal to

$$\frac{\Gamma_r(a+b)}{\Gamma_r(a)\Gamma_r(b)} (\det x)^{a-(r+1)/2} \int_{\mathscr{S}_r^+} \exp[-\mathrm{Tr}((x + \mathbb{1}_r)y)] \frac{(\det y)^{a+b-(r+1)/2}}{\Gamma_r(a+b)} dy. \tag{2}$$

Note that x belongs to $\mathscr{S}_r^+$ and then $(\mathbb{1}_r + x)$ also belongs to $\mathscr{S}_r^+$. Therefore, using Theorem 8.1 we can evaluate the integral which appears in (2) and obtain for the density of X

$$\frac{\Gamma_r(a+b)}{\Gamma_r(a)\Gamma_r(b)}(\det x)^{a-(r+1)/2}[\det(\mathbb{1}_r + x)]^{-a-b}.$$ ∎

Corollary 1 *Let X be the random variable defined in Theorem 1. Then the random matrix*

$$U = X(\mathbb{1}_r + X)^{-1}$$

has a distribution, which is designated by $B_r'(a, b)$, whose probability density function is defined by

$$\frac{\Gamma_r(a+b)}{\Gamma_r(a)\Gamma_r(b)}(\det u)^{a-(r+1)/2}[\det(\mathbb{1}_r - u)]^{b-(r+1)/2},$$

where $u \in \mathscr{S}_r^+$ and $(\mathbb{1}_r - u) \in \mathscr{S}_r^+$.

Proof The transformation

$$u = x(\mathbb{1}_r + x)^{-1}, \qquad x = (\mathbb{1}_r - u)^{-1}u \tag{3}$$

is a one-to-one differentiable mapping between the matrix $x \in \mathscr{S}_r^+$ and the matrix u where u and $\mathbb{1}_r - u$ belong to $\mathscr{S}_r^+$. Then Theorem 3.4 shows that the Jacobian of (3) is given by

$$\left|\frac{\partial x}{\partial u}\right| = |\det(\mathbb{1}_r - u)|^{-r-1}.$$

The proof is completed by using Theorem 1. ∎

Corollary 2 *Let U be a random matrix having the $\Gamma_r(a, \mathbb{1}_r)$ distribution and let V be a random matrix which is independent of U. If the random matrix $({}^tVV)^{-1}$ has the $\Gamma_r(b, \mathbb{1}_r)$ distribution, and if*

$$2a + 1 > r, \qquad 2b + 1 > r,$$

then ${}^tVUV$ has the $B_r(a, b)$ distribution.

Proof Use Theorem 6.3 to show that the conditional distribution of ${}^tVUV$ given V is the $\Gamma_r(a, {}^tVV)$ distribution. Then, apply Theorem 1 to $Y = ({}^tVV)^{-1}$. ∎

Theorem 2 *Let Y be a random matrix having the $\Gamma_r(b, \mathbb{1}_r)$ distribution and let X be a random matrix which is independent of Y and almost everywhere nonsingular. If the random matrix $X\,{}^tX$ has the $\Gamma_r(a, \mathbb{1}_r)$ distribution and if*

$2a+1>r$, $2b+1>r$, *then the random matrix* $XY^{-1}\,{}^tX$ *has the* $B_r(a,b)$ *distribution.*

Proof First, use Theorem 3.4 to prove the following:

Lemma *Let* X *be a random matrix having the* $B_r(a,b)$ *distribution; then* X^{-1} *has the* $B_r(b,a)$ *distribution.*

Then complete the proof by applying Corollary 2 and noting that

$$(XY^{-1}\,{}^tX)^{-1} = {}^tX^{-1}YX^{-1}. \qquad \blacksquare$$

Remark Bartlett's decomposition or that of Exercise 18 shows that there exists a matrix X such that $X\,{}^tX$ has the $\Gamma_r(a, \mathbb{1}_r)$ distribution. Moreover, if $2a+1>r$, there exists a nonsingular random matrix $Y[r,r]$ such that both $Y\,{}^tY$ and ${}^tYY$ have $\Gamma_r(a, \mathbb{1}_r)$ distributions.

Theorem 3 *Let* X *be a random matrix having the* $B_r(a,b)$ *distribution. Then* $b_r(a,b)$ *the distribution of* $R(X)$ *defined by* (1.4) *has probability density function*

$$\frac{\pi^{r^2/2}\Gamma_r(a+b)}{\Gamma_r(a)\Gamma_r(b)\Gamma_r(r/2)}\left(\prod_1^r u_i\right)^{a-(r+1)/2}\left(\prod_1^r (1+u_i)\right)^{-a-b}\prod_{1\leqslant i<j\leqslant r}(u_i-u_j),$$

where $u_1>u_2>\cdots>u_r>0$.

Corollary 3 *Let* U *be a random matrix having the* $B_r(a,b)$ *distribution. Then the distribution of the random variable* $\det U$ *is that of the product of* r *independent random variables having distributions*

$$\beta(a-\tfrac{1}{2}j,\, b-\tfrac{1}{2}j), \qquad j=0\cdots r-1.$$

Proof Using Theorem 2, we have

$$U = XY^{-1}\,{}^tX$$

and then

$$\det U = \det(X\,{}^tX)\det Y^{-1} = \det(X\,{}^tX)/\det Y.$$

The proof is completed by using Corollary 7.2. $\blacksquare$

Theorem 4 *Let* U *be a random matrix having the* $B'_r(a,b)$ *distribution, then the random variable* $\det U$ *has the distribution of the product of* r *independent variables having distributions*

$$\beta'(a-\tfrac{1}{2}j,\, b), \qquad j=0,\ldots,r-1.$$

Proof Using Corollary 1, we see that for every positive t

$$E(\det U)^t = \frac{\Gamma_r(a+b)}{\Gamma_r(a+b+t)} \frac{\Gamma_r(a+t)}{\Gamma_r(a)}.$$

However, this quantity is equal to

$$\prod_{j=0}^{r-1} \frac{\Gamma(a+b-\frac{1}{2}j)\Gamma(a+t-\frac{1}{2}j)}{\Gamma(a+b+t-\frac{1}{2}j)\Gamma(a-\frac{1}{2}j)}. \tag{4}$$

Now, if we consider r independent random variables, $V_0, \ldots, V_{r-1}$ having $\beta'(a - \frac{1}{2}j, b)$ $j = 0, \ldots, r-1$ distributions, an elementary computation shows that for every positive t the expectation

$$E\left[\left(\prod_{j=0}^{r-1} V_j\right)^t\right]$$

is equal to (4). It is obvious that this last condition characterizes the distribution of $\det U$. ∎

Theorem 5 *Let X be a random matrix having the $N(O, \mathbb{1}_n \otimes \mathbb{1}_r)$ distribution $(r \geqslant n)$ and let Y be a random matrix which is independent of X and which has the $\Gamma_r(b, 2\mathbb{1}_r)$ distribution, $(2b+1 > r)$; then the random matrix $XY^{-1}\,{}^tX$ has the $B_n(r/2, b - \frac{1}{2}(r-n))$ distribution.*

This theorem follows directly from Exercise 20. Note that Theorems 2 and 5 solve the same problem but under different conditions. See Exercise 21 for a statistical application of Theorem 5.

Theorem 6 *Let U and V be two independent random matrices having $\Gamma_r(a, \Lambda)$ and $\Gamma_r(b, \Lambda)$ distributions respectively. If Λ is nonsingular and if $2a + 1 > r$, then the distribution of the random variable*

$$Z = \frac{\det U}{\det(U+V)}$$

is the distribution of a product of independent random variables having distributions

(i) $\beta'(a - \frac{1}{2}j, b)$, $\quad j = 0, \ldots, r-1$, *if* $2b + 1 > r$,
(ii) $\beta'(a - \frac{1}{2}(r-j), r/2)$, $\quad j = 1, \ldots, s$, *if* $2b$ *is an integer* $s \leqslant r$.

Proof Using Theorem 6.3 with a matrix A such that $\Lambda = A\,{}^tA$ we reduce the theorem to the case $\Lambda = \mathbb{1}_r$ for (i) and to the case $\Lambda = 2\mathbb{1}_r$ for (ii). We also have $Z^{-1} = \det(\mathbb{1}_r + VU^{-1})$.

In case (i), there exists a nonsingular random matrix X such that

$$V = {}^{t}XX, \tag{5}$$

where $X\,{}^{t}X$ also has the $\Gamma_r(b, \mathbb{1}_r)$ distribution. Thus we have

$$Z^{-1} = \det(\mathbb{1}_r + XU^{-1}\,{}^{t}X). \tag{6}$$

Using Theorem 2, we see that $XU^{-1}\,{}^{t}X$ has the $B_r(b, a)$ distribution and then $[\mathbb{1}_r + XU^{-1}\,{}^{t}X]^{-1}$ has the $B_r'(a, b)$ distribution. The proof is completed by applying Theorem 4.

In case (ii), let $X[s, r]$ be a random matrix having the $N(O, \mathbb{1}_s \otimes \mathbb{1}_r)$ distribution. Using Theorem 6.1 we see that formula (5) holds. Using Theorem 3.7 we see that formula (6) remains valid if we change $\mathbb{1}_r$ to $\mathbb{1}_s$. Then, using Theorem 5 we see that $XU^{-1}\,{}^{t}X$ has the $B_s(r/2, a - \frac{1}{2}(r - s))$ distribution. The proof is completed by the same argument as for (i). ∎

Lemma 1 *Let X and Y be two independent random matrices having $\Gamma_r(a, \mathbb{1}_r)$ and $\Gamma_r(b, \mathbb{1}_r)$ distributions respectively $(2a + 1 > r, 2b + 1 > r)$; then $R(XY^{-1})$ has the $b_r(a, b)$ distribution.*

Proof The joint probability density function of X and Y is

$$\frac{1}{\Gamma_r(a)\Gamma_r(b)} \exp[-\mathrm{Tr}(x + y)](\det x)^{a-(r+1)/2}(\det y)^{b-(r+1)/2}.$$

Performing the transformations defined in Theorem 3.6 we get

$$\frac{2^r}{\Gamma_r(a)\Gamma_r(b)} \exp[-\mathrm{Tr}\,{}^{t}EE](\det E)^{2a+2b-r}(\det F)^{a-(r+1)/2}$$
$$\times [\det(\mathbb{1}_r - F)]^{b-(r+1)/2} \prod_{1 \leqslant i < j \leqslant r} (f_i - f_j),$$

where the first column of E is positive and F is diagonal. The function of E to be integrated is invariant under change of sign of elements in the first column. Thus we may simultaneously delete the constant 2^r and ignore the condition that E have a positive first column. Moreover, this function depends only on ${}^{t}EE$ which is equal to $X + Y$; using Exercise 18, we can perform the integration with respect to E. Changing f_i to $\lambda_i(1 + \lambda_i)^{-1}$ completes the proof. (See Exercise 11.) ∎

Theorem 7 *Let U be a random matrix having the $B_r(a, r/2)$ distribution; then $R(U)$ has the $b_r(a, r/2)$ distribution.*

Proof Let Z be a random matrix having the $N(O, \mathbb{1}_r \otimes \mathbb{1}_r)$ distribution and let X be a random matrix which has the $\Gamma_r(a, 2\mathbb{1}_r)$ distribution and is independent of Z. Corollary 2 shows that the random matrix

$$U = {}^t Z^{-1} X Z^{-1}$$

has the $B_r(a, r/2)$ distribution. On the other hand, the two matrices U and $X({}^t ZZ)^{-1}$ have the same eigenvalues and ${}^t ZZ$ has the $\Gamma_r(r/2, \mathbb{1}_r)$ distribution. Thus the previous theorem can be applied. ∎

10. Generalized Noncentral Gamma Distributions

Definition 1 *Let r be a positive integer, and a be a positive number. Let K and Λ be two matrices belonging to $\mathscr{S}_r^+$. When it exists, we designate by $\Gamma_r(a, K, \Lambda)$ the distribution on $\mathscr{S}_r^+$ having the characteristic function*

$$[\det(\mathbb{1}_r - iT\Lambda)]^{-a} \exp\{i \operatorname{Tr}(K(\mathbb{1}_r - iT\Lambda)^{-1} T)\}, \qquad T \in \mathscr{S}_r^+.$$

The following theorem may be proved by the same argument as used in Theorem 6.1.

Theorem 1 *Let $X[n, r]$ be a random matrix having the $N(m, \mathbb{1}_n \otimes \Lambda)$ distribution, then the random matrix ${}^t XX$ has the $\Gamma_r(n/2, {}^t mm, 2\Lambda)$ distribution.*

Theorem 2 $\Gamma_r(a, K, \Lambda) * \Gamma_r(b, K', \Lambda) = \Gamma_r(a + b, K + K', \Lambda)$.

Theorem 3 *Let X be a random matrix having the $\Gamma_r(a, K, \Lambda)$ distribution and let $A[r, m]$ be a given matrix. Then the random matrix ${}^t AXA$ has the $\Gamma_m(a, {}^t AKA, {}^t A\Lambda A)$ distribution.*

Proof The proof is analogous to that of Theorem 6.3. We let $Y = {}^t AXA$ and then have $\Phi_Y(T) = \Phi_X(AT\,{}^t A)$. Now,

$$\Phi_Y(T) = [\det(\mathbb{1}_m - iT\,{}^t A\Lambda A)]^{-a} \exp\{i \operatorname{Tr}(K(\mathbb{1}_r - iAT\,{}^t A\Lambda)^{-1} AT\,{}^t A)\}.$$

But,

$$\operatorname{Tr}[K(\mathbb{1}_r - iAT\,{}^t A\Lambda)^{-1} AT\,{}^t A] = \operatorname{Tr}[{}^t AK(\mathbb{1}_r - iAT\,{}^t A)^{-1} AT].$$

The proof is completed by noting that

$$(\mathbb{1}_r - iAT\,{}^t A\Lambda)^{-1} A = A(\mathbb{1}_m - iT\,{}^t A\Lambda A)^{-1}.$$ ∎

Corollary *Let X be a random matrix having the $\Gamma_r(a, \Gamma, \Lambda)$ distribution where Λ is nonsingular. Then there exists a matrix A such that the random* *matrix* $Y = {}^tAXA$ *has the* $\Gamma_r(a, K, \mathbb{1}_r)$ *distribution, where* K *is the diagonal matrix whose diagonal elements are the eigenvalues of* $\Gamma\Lambda^{-1}$.

Proof Lemma 7.2.1 shows that there exists a nonsingular matrix A such that $\Lambda^{-1} = A\,{}^tA$, ${}^tA\Gamma A = K$, where the diagonal elements of the diagonal matrix K are the eigenvalues of $\Gamma\Lambda^{-1}$. Apply Theorem 3 to complete the proof. ∎

Theorem 4 *Let* X *be a random matrix having the* $\Gamma_r(a, K, \Lambda)$ *distribution. If* X, K, Λ *are partitioned in blocks of the same order*

$$X = \begin{pmatrix} X' & X'' \\ {}^tX'' & X''' \end{pmatrix}, \qquad K = \begin{pmatrix} K' & K'' \\ {}^tK'' & K''' \end{pmatrix}, \qquad \Lambda = \begin{pmatrix} \Lambda' & \Lambda'' \\ {}^t\Lambda'' & \Lambda''' \end{pmatrix},$$

where X', K', Λ' *are of order* r', *then* X' *and* X''' *have the* $\Gamma_{r'}(a, K', \Lambda')$ *and* $\Gamma_{r-r'}(a, K''', \Lambda''')$ *distribution, respectively. Moreover if*

$$\Lambda'' = 0, \qquad K'' = 0,$$

X' *and* X''' *are independent.*

Proof The proof is similar to that of Theorem 6.4 with the addition of the computation,

$$\mathrm{Tr}(K(\mathbb{1}_r - iT\Lambda)^{-1}T) = \mathrm{Tr}(\mathbb{1}_r - iT\Lambda)^{-1}TK).$$

The proof is completed by noting that

$$(\mathbb{1}_r - iT\Lambda)^{-1} = \begin{pmatrix} (\mathbb{1}_r - iT'\Lambda')^{-1} & 0 \\ 0 & (\mathbb{1}_{r-r'} - iT'''\Lambda''')^{-1} \end{pmatrix}$$

and

$$TK = \begin{pmatrix} T'K' & 0 \\ 0 & T'''K''' \end{pmatrix}.$$ ∎

Theorem 5 *Using the partitions of Theorem 6.4, assume that* Λ' *is nonsingular. Then the conditional distribution of* X''' *given* X' *is the* $\Gamma_{r-r'}(a, {}^t\Lambda''\Lambda'^{-1}X'\Lambda'^{-1}\Lambda'', \Lambda''' - {}^t\Lambda''\Lambda'^{-1}\Lambda'')$ *distribution.*

Proof Let $A = -{}^t\Lambda''\Lambda'^{-1}$, $B = \Lambda''' - {}^t\Lambda''\Lambda'^{-1}\Lambda''$. Assume that X' has the $\Gamma_{r'}(a, \Lambda')$ distribution and that the conditional distribution of X''' given X' is that given by the theorem. Then, the joint characteristic function of X' and X''' is given by

$$(T', T''') \to E_{X'}(\Phi_{X'''}^{X'}(T''')\exp[i\,\mathrm{Tr}(T'X')]) \tag{1}$$

where $\Phi_{X'''}^{X'}$ is the conditional characteristic function of X''' given X'. Using

Definition 1, the function (1) becomes

$$[\det(\mathbb{1}_{r-r'} - iT'''B)]^{-a}\Phi_{X'}(T' + {}^{t}A(\mathbb{1}_{r-r'} - iT'''B)^{-1}T'''A).$$

Using Definition 6.1 this is equal to

$$[\det(\mathbb{1}_{r-r'} - iT'''B)]^{-a} \det[\mathbb{1}_{r'} - iT'\Lambda' - i\,{}^{t}A(\mathbb{1}_{r-r'} - iT'''B)^{-1}T'''A\Lambda']^{-a}. \quad (2)$$

The following computation shows that (2) is the marginal characteristic function of (X''', X') when we assume that

$$\begin{pmatrix} X' & X'' \\ X'' & X''' \end{pmatrix}$$

has the $\Gamma_r(a, \Lambda)$ distribution. We have, using (7.4)

$$\det(\mathbb{1}_{r-r'} - iT'''B) \det[\mathbb{1}_{r'} - iT'\Lambda' - i\,{}^{t}A(\mathbb{1}_{r-r'} - iT'''B)^{-1}T'''A\Lambda']$$

$$= \det\begin{pmatrix} \mathbb{1}_{r-r'} - iT'''B & iT'''A\Lambda' \\ {}^{t}A & \mathbb{1}_{r'} - iT'\Lambda' \end{pmatrix} \quad (3)$$

and this last matrix is equal to

$$\begin{pmatrix} \mathbb{1}_{r-r'} & 0 \\ {}^{t}A & \mathbb{1}_{r'} \end{pmatrix} - i\begin{pmatrix} T''' & 0 \\ 0 & T' \end{pmatrix}\begin{pmatrix} B & -A\Lambda' \\ 0 & \Lambda' \end{pmatrix}.$$

The proof is completed by noting that

$$\begin{pmatrix} \mathbb{1}_{r-r'} & 0 \\ {}^{t}A & \mathbb{1}_{r'} \end{pmatrix}^{-1} = \begin{pmatrix} \mathbb{1}_{r-r'} & 0 \\ -{}^{t}A & \mathbb{1}_{r'} \end{pmatrix},$$

and then (3) is equal to the determinant of

$$\begin{pmatrix} \mathbb{1}_{r-r'} & 0 \\ 0 & \mathbb{1}_{r'} \end{pmatrix} - i\begin{pmatrix} T''' & 0 \\ 0 & T' \end{pmatrix}\begin{pmatrix} B & -A\Lambda' \\ 0 & \Lambda' \end{pmatrix}\begin{pmatrix} \mathbb{1}_{r-r'} & 0 \\ -{}^{t}A & \mathbb{1}_{r'} \end{pmatrix}.$$

∎

Theorem 6 *Let a be a real number and let q be an integer such that*

$$2a + 1 > q > 0.$$

When it exists, we designate by g_a^q the function on $\mathcal{S}_q^+$ where one of the two equivalent conditions is fulfilled.

(i) *$g_a^q(x)$ depends only on the eigenvalues of x and for every diagonal matrix Λ of order q we have*

$$E[g_a^q(X)] = \exp[\operatorname{Tr} \Lambda],$$

where X is a random matrix having the $\Gamma_q(a, \Lambda)$ distribution.

(ii) *for every matrix* Λ *belonging to* $\mathcal{S}_q^+$ *we have*

$$E[g_a^q(X)] = \exp[\operatorname{Tr} \Lambda], \tag{4}$$

where X *is a random matrix having the* $\Gamma_q(a, \Lambda)$ *distribution.*

Proof Theorem 8.2 shows that (4) has, at most, only one solution. Let h_a^q be this solution when it exists. For every orthogonal matrix C, the random matrix ${}^tCXC$ has the $\Gamma_q(a, {}^tC\Lambda C)$ distribution where X is a random matrix having the $\Gamma_q(a, \Lambda)$ distribution. Then we have

$$E[h_a^q({}^tCXC)] = \exp[\operatorname{Tr}({}^tC\Lambda C)] = \exp[\operatorname{Tr} \Lambda].$$

This implies that $h_a^q(x)$ and $h_a^q({}^tCxC)$ are two solutions of (4) and thus are equal, i.e., $h_a^q(x)$ depends only on the eigenvalues of x. Conversely, we assume that the function g_a^q exists, and we recall that every matrix Λ can be written as

$$\Lambda = {}^tC\Lambda_0 C,$$

where C is an orthogonal matrix and Λ_0 a diagonal one. Let X be a random matrix having the $\Gamma_q(a, \Lambda_0)$ distribution. By assumption, we have

$$E[g_a^q(X)] = \exp[\operatorname{Tr} \Lambda_0].$$

But

$$g_a^q({}^tCXC) = g_a^q(X).$$

Then

$$E[g_a^q({}^tCXC)] = \exp[\operatorname{Tr} \Lambda_0] = \exp[\operatorname{Tr} {}^tC\Lambda_0 C] = \exp[\operatorname{Tr} \Lambda].$$

Thus g_a^q satisfies (4) since the random matrix ${}^tCXC$ has the $\Gamma_q(a, \Lambda)$ distribution. ∎

Theorem 7 *Let* $h_a^q(u_1, \ldots, u_q)$ *be the function defined by* $h_a^q(U(x)) = g_a^q(x)$ $\forall x \in \mathcal{S}_q^+$, *where* g_a^q *is defined in Theorem 6 and* $U(x)$ *is the sequence of the fundamental symmetric functions of the eigenvalues of* x. *Then we have*

$$\forall q' < q, \qquad h_a^{q'}(u_1, \ldots, u_{q'}) = h_a^q(u_1, \ldots, u_{q'}, 0, \ldots, 0).$$

Proof We recall (see Exercise 5) that the sequence $U(x) = (U_1(x), \ldots, U_q(x))$ is defined by

$$\forall s \in \mathbb{R}, \qquad \det(X - s\mathbb{1}_q) = \sum_{i=0}^{q} (-s)^{q-i} U_i(X). \tag{5}$$

Let x' be a square matrix of order q' ($q' < q$). Using (5) we see that

$$U\left(\begin{pmatrix} x' & 0 \\ 0 & 0 \end{pmatrix}\right) = (U(x'), 0, \ldots, 0). \tag{6}$$

Let Λ be the matrix,

$$\Lambda = \begin{pmatrix} \Lambda_0 & 0 \\ 0 & 0 \end{pmatrix},$$

where Λ_0 is a matrix belonging to $\mathscr{S}_{q'}^+$. We see easily that if X, a random matrix has the $\Gamma_q(a, \Lambda)$ distribution, then

$$X = \begin{pmatrix} X_0 & 0 \\ 0 & 0 \end{pmatrix},$$

where X_0 is a random matrix having the $\Gamma_{q'}(a, \Lambda_0)$ distribution.

Using (6) we have

$$U(X) = (U(X_0), 0, \ldots, 0),$$

and by assumption

$$E[h_a^q(U(X_0), 0, \ldots, 0)] = \exp[\operatorname{Tr} \Lambda] = \exp[\operatorname{Tr} \Lambda_0]. \quad \blacksquare$$

Remark This theorem shows that, if $2a$ is an integer, we can easily compute the functions h_a^q with the help of the function h_a^{2a}. However the function h_a^{2a} can be computed with the help of the generating function of the Haar invariant distribution (see Exercises 12 and 25). Thus, the function g_a^q exists when $2a$ is an integer greater than or equal to q.

Theorem 8 *Let X be a random matrix having the $\Gamma_r(a, K, \mathbb{1}_r)$ distribution, where*

$$K = \begin{pmatrix} K' & 0 \\ 0 & 0 \end{pmatrix}$$

and the matrix K' is a nonsingular diagonal matrix of order q ($q \leqslant r$). If the function g_a^q exists, then the probability density function of X is equal to

$$x \to \exp[-\operatorname{Tr} K] \exp[-\operatorname{Tr} x] \frac{(\det x)^{a - \frac{1}{2}(r+1)}}{\Gamma_r(a)} g_a^q(K'x'), \tag{7}$$

where x' is the submatrix of x obtained by deleting the last $r - q$ rows and columns.

Proof Compute $L(S)$, the Laplace transform of (7),

$$\forall S \in \mathscr{S}_r^+, \qquad L(S) = \exp[-\operatorname{Tr} K][\det(\mathbb{1}_r + S)]^{-a} E[g_a^q(K'X')],$$

where X is a random matrix having the $\Gamma_r(a, (S + \mathbb{1}_r)^{-1})$ distribution. Then X' has the $\Gamma_q(a, ((S + \mathbb{1}_r)^{-1})')$ distribution and thus, $K'^{1/2}X'K'^{1/2}$ has the

$\Gamma_q(a, K'^{1/2}[(S + \mathbb{1}_r)^{-1}]'K'^{1/2})$ distribution. Use the properties of the function g_a^q to obtain

$$g_a^q(K'X') = g_a^q(K'^{1/2}X'K'^{1/2})$$

and

$$L(S) = \exp[-\operatorname{Tr} K][\det(\mathbb{1}_r + S)]^{-a} \exp\{\operatorname{Tr}(K'^{1/2}[(S + \mathbb{1}_r)^{-1}]'K'^{1/2})\}.$$

Now an easy computation yields

$$L(S) = [\det(\mathbb{1}_r + S)]^{-a} \exp[-\operatorname{Tr}(K(S + \mathbb{1}_r)^{-1}S)].$$

Formally change S to $-iT$ to obtain the characteristic function of the $\Gamma_r(a, K, \mathbb{1}_r)$ distribution. ∎

Exercises

1. Let A and B be two matrices of the same order. Show that

$$\operatorname{Tr}(A + B) = \operatorname{Tr}(A) + \operatorname{Tr}(B).$$

If, moreover, A and ${}^tB$ have the same order, show that

$$\operatorname{Tr}(AB) = \operatorname{Tr}(BA).$$

2. Using the definition of the Kronecker product of two matrices, prove Properties (1.1), (1.2), and (1.3).

3. Let X be an orthogonal matrix which has the Haar invariant distribution H_r. Show that the signs of the elements in the first column are independent, identically distributed random variables with distribution: plus and minus, each with probability $\frac{1}{2}$. What is the conditional distribution of X given that its first column is positive?

4. Show that the conditional Haar distribution on $\mathscr{C}_r^+$ is the unique distribution on $\mathscr{C}_r^+$ with the property: if X is a random matrix with this distribution, then for every orthogonal matrix C of order r, the matrix obtained from XC by multiplying each row by the sign of the first element of the row has the same distribution as X.

5. Show by induction that the Jacobian of the transformation

$$(x_1 \geqslant x_2 \geqslant \cdots \geqslant x_n) \rightleftharpoons \begin{pmatrix} u_1 = x_1 + \cdots + x_n \\ u_2 = x_1x_2 + \cdots + x_{n-1}x_n \\ \vdots \\ u_n = x_1x_2 \cdots x_n \end{pmatrix}$$

is equal to

$$\prod_{1 \leqslant i < j \leqslant n} (x_i - x_j).$$

Let X be a square matrix of order r, and let $U(X) = \{U_1(X), \ldots, U_r(X)\}$ be the sequence defined by

$$\forall s \in \mathbb{R}, \qquad \det(X - s\mathbb{1}_r) = \sum_{i=0}^{r} (-s)^{r-i} U_i(X).$$

Write $U(X)$ as a function of $R(X)$, defined in (1.4). If X is a random matrix and has the $\Gamma_r(a, \mathbb{1}_r)$ distribution, find the probability distribution function of $U(X)$. If X has the $B_r(a, b)$ distribution, find the probability distribution function of $U(X)$. If X has the $B'_r(a, b)$ distribution find the probability distribution functions of $R(X)$ and of $U(X)$.

6. Let U be a random matrix whose rows are independent and let Λ_i be the quadratic form defined by the covariance of the ith row of U. Show that

$$\Lambda_U(x) = \sum_i \Lambda_i(x_i),$$

where x_i is the ith row of x.

7. Let U be a random matrix which has covariance Λ_U. Show that

$$\Lambda_{{}^tU}(x) = \Lambda_U({}^tx)$$

and that the general term of $\Lambda_{\tilde{U}}$ is equal to the covariance between U_{ij} and $U_{i'j'}$.

8. Prove Theorem 2.2.

9. Prove the implications (2.1) $\Leftrightarrow$ (2.2) and (2.2) $\Leftrightarrow$ (2.3) and show that $\Lambda_{{}^t\tilde{U}}$ is the Kronecker (or direct) product of Λ' and Λ''.

10. Prove Theorem 3.5 by means of induction on n.

11. Prove the first part of Theorem 3.6, i.e., the existence of matrices A and B. Prove that the eigenvalues of AB^{-1} equal $f_i/(1 - f_j)$ $(j = 1, \ldots, r)$.

12. Let C be an orthogonal random matrix of order r having the Haar invariant distribution. Show that there exists a function g_r, such that for every nonsingular matrix S of order r,

$$E(\exp \operatorname{Tr}(CS)) = g_r(R(S\,{}^tS)).$$

Compute g_2.

13. Let $X[n, r]$ be a random matrix having the $N(O, \Lambda \otimes \Lambda')$ distribution and let $A[r, n]$ be a given matrix. Show that

$$E(\exp \operatorname{Tr}(XA)) = \exp[\tfrac{1}{2} \operatorname{Tr}({}^tA\Lambda' A\Lambda)].$$

14. Let X be a random matrix having the $\Gamma_r(a, \Lambda)$ distribution and let x be a given vector of $\mathbb{R}^r$. Show that the random variable ${}^t x X x$ has the $\Gamma(a, {}^t x \Lambda x)$ distribution. Furthermore, let $x_1, \ldots, x_k$ be k vectors in $\mathbb{R}^r$, which are orthogonal for the inner product defined by Λ^{-1}, show that the random variables ${}^t x_j X x_j$ $(j = 1, \ldots, k)$ are independent.

15. Let X be a random matrix having the $\Gamma_r(a, \Lambda)$ distribution. Show that

$$E(X) = a\Lambda.$$

16. Let δ and p $(1 \leqslant \delta \leqslant p)$ be two given integers and let $\{X_j^i; i = 1, \ldots, p, j = \delta, \ldots, p, j \geqslant i\}$ be independent random variables, where

$$X_j^i \text{ has the } N(O, \lambda_i/2) \text{ distribution, if } j > i$$

$$(X_i^i)^2 \text{ has the } \Gamma(a + \tfrac{1}{2}(j - p), \lambda_i) \text{ distribution,}$$

and λ_i $(i = 1, \ldots, p)$ are given positive numbers. Denote by X the triangular matrix having X_j^i as the general term (and zero elsewhere). Find the characteristic function of $X\,{}^tX$. Investigate the special cases, where $\delta = 1$ and where $2a$ is an integer.

17. Let X be a random matrix having the nonsingular $\Gamma_r(a, \Lambda)$ distribution and let A be a given symmetric matrix of order r. For all real numbers t, s $(s > 0)$ compute

$$E(\exp[it \operatorname{Tr}(AX)](\det X)^s).$$

Show that this computation gives the joint distribution of $\det X$ and $\operatorname{Tr} AX$. Find the marginal distribution of each of these random variables. In your opinion, which is the most interesting case?

18. Let a be a real number and let n and r be two integers, such that

$$2a + 1 > r, \qquad n \geqslant r > 0.$$

Show the existence of a random matrix $Y[r, n]$, having the probability density function,

$$\frac{\Gamma_r(n/2)}{\Gamma_r(a)} \pi^{-nr/2} \exp[-\operatorname{Tr}(y\,{}^t y)][\det(y\,{}^t y)]^{a-(n/2)}.$$

Find the distribution of $Y\,{}^tY$.

19. Let $X = \{X_{ij}\}$ be a random matrix having the $\Gamma_r(a, \mathbb{1}_r)$ distribution. Find the joint distribution of the random variables,

$$R_{ij} = \frac{X_{ij}}{(X_{ii}X_{jj})^{1/2}}, \qquad i \neq j = 1, \ldots, r.$$

20. Let Z be a random matrix having the $N(O, \mathbb{1}_n \otimes \mathbb{1}_r)$ distribution and let $Y[r, r]$ be a random matrix which is independent of Z, such that the random matrix $Y\,{}^tY$ has the $\Gamma_r(b, 2\mathbb{1}_r)$ distribution, where $2b + 1 > r$. Determine the probability density function of $X = ZY^{-1}$ and find the distribution of $X\,{}^tX$ and that of ${}^tXX$.

21. (Hotelling) Let $\bar{X}$ and S be as defined in the corollary to Theorem 6.1. Show that the random variable $T = {}^t(\bar{X} - m)S^{-1}(\bar{X} - m)$, has the $\beta(r/2, (n - r)/2)$ distribution.

22. Let X be a random matrix having the $\Gamma_r(a, K, \Lambda)$ distribution. What is the distribution of ${}^txXx$, where x is a given vector of $\mathbb{R}^r$? Deduce from this that

$$E(X) = a\Lambda + K.$$

23. Let U and V be two independent matrices having the $\Gamma_r(a, \mathbb{1}_r)$ and the $\Gamma_r(b, \mathbb{1}_r)$ distributions, respectively, where $2a + 1 > r$. Show that for every positive number t, the conditional expectation $E((\det(U + V)^{-t} | U)$ does not depend on a. Denote this conditional expectation by $f(U)$ and compute the integral, $\int_{\mathscr{S}_r^+} f \, d\Gamma_r(a + t, \mathbb{1}_r)$. Use this result to compute, for every positive t,

$$E\left[\frac{\det U}{\det(U + V)}\right]^t$$

and show that this last result gives a direct proof of Theorem 9.6.

24. Using Condition (i) of Theorem 10.6, show that the functions h_a^1 and h_a^2 defined in Theorem 10.7 are given by

$$h_a^1(x) = \Gamma(a) \sum_{k=0}^{\infty} \frac{x^k}{k!\Gamma(a + k)}$$

$$\pi^{1/2} h_a^2(x, y) = \Gamma_2(a) \sum_{k, k' \geqslant 0} \frac{x^k y^{k'}}{k!k'!\Gamma(a + k' - \frac{1}{2})\Gamma(a + 2k' + k)}.$$

25. Using the notation of Theorem 10.7, for every integer r compute the function $h_{r/2}^r$ with the help of the function g_r found in Exercise 12.

26. Let $X[n, r]$ be a random matrix having the $N(M, \mathbb{1}_n \otimes \Lambda)$ distribution and C an orthogonal random matrix of order n which has the Haar invariant distribution and is independent of X. Find the probability density function of the random matrix $Y = CX$ and deduce from this the probability density function of ${}^tYY$. Apply this result to obtain the probability density function of a noncentral gamma distribution.

27. Consider a sample of size $n > 2$ from the bivariate normal distribution. Show that the sample correlation coefficient has the following probability density function on [0, 1]

$$f(r) = \frac{2^{n-3}(1-\rho^2)^{(n-1)/2}}{\pi(n-3)!}(1-r^2)^{(n-4)/2}\sum_{j=0}^{\infty}\frac{(2\rho r)^j}{\rho!}\left[\Gamma\left(\frac{n+j-1}{2}\right)\right]^2$$

where ρ is the correlation coefficient.

28. Let U be a random matrix having the $B_r(a, b)$ distribution and let t be a real number, where $0 < 2t < 2b + 1 - r$. Compute $E(\det U)^t$ and thus give a new proof of Corollary 9.3.

29. Prove Theorem 9.2 directly without using Theorem 9.1 and the lemma following Theorem 9.2.

CHAPTER 9

Linear-Normal Statistical Spaces

Classical linear models and multivariate analysis are based on statistical spaces studied in this chapter. We emphasize the geometrical form of the basic ideas so that the final formulas involving matrices appear as simple applications of previous general ideas. The reader will note that the hypotheses which are made on the mean in these statistical spaces are always linear and that the covariance is always the product of a known term and an unknown term. Moreover, all the linear spaces that we consider are defined over the field of real numbers. More precisely, in the same way as the foundations (from a stochastic point of view) of multivariate analysis have been built up in Chapter 8, the foundations (from a statistical point of view) are being defined in this chapter. The guiding principle uses the geometrical language of Hilbert spaces (of finite dimension). The results of Dieudonné [13], which the reader is invited to consult systematically, will be used frequently.

In Section 1, we recall some useful properties of quadratic forms, e.g., the Cochran theorem. Here the notion of covariance as a quadratic form is to be distinguished from that of the covariance matrix representing the covariance in a given basis. Then one can see easily that the classical linear models are defined on the statistical space defined in Section 2. In Section 3 a fundamental property of projections is established for normal distributions. The methods of statistical analysis of linear models are then expressed very simply in Sections 4 and 5 and are applied (Sections 6–9) to the case of analysis of variance in experimental design. Without further difficulty greater generality will be attained than in the usual representations. Finally, using techniques developed in Chapter 8, Sections 2 and 3 are generalized in Section 10, yielding the foundations of multivariate analysis.

To conclude, the aim of this chapter is to obtain a more elegant and more rapid presentation of the classical theory of linear models and to show that multivariate analysis is constructed on the same basic ideas, which are simple and few. Exercises 5, 12, 13, 15, 16, and 26 will help the reader to see the bearing of such a presentation, in some simple examples.

1. The Cochran Theorem

Theorem 1 *Let q be a nonnegative definite quadratic form on the finite-dimensional linear space E. Then there exists a unique subspace E_0 and a subspace F such that*

(1) $E = E_0 + F$;
(2) $q(x) = 0, \quad x \in F \Rightarrow x = 0$;
(3) $q(x) = 0, \quad \forall x \in E_0$;
(4) E_0 *and F are orthogonal with respect to the inner product defined by q.*

We call E_0 the *kernel* of q and F the *support* of E. Moreover, we have $\operatorname{rank}(q) = \dim E - \dim E_0$, where "dim" means "dimension of." This theorem is equivalent to:

Theorem 1′ *Every vector x in E can be written in the following unique way:*

$$x = x_0 + y, \qquad x_0 \in E_0, \quad y \in F,$$

where

(1′) $q(x) = q(y)$,
(2′) $q(x_0) = 0$,
(3′) $\varphi(x_0, y) = 0$,

and φ is the bilinear form associated with q, i.e.,

$$2\varphi(x, y) = q(x + y) - q(x) - q(y).$$

Thus it can be seen that (1′) and (2′) imply (3′).

Lemma *If a basis has been chosen for E, then $q(x) = {}^t xAx$ where A is a matrix belonging to $\mathcal{S}_n^+$ ($n = \dim E$) and the following implication is true*

$$q(x) = 0 \quad \Rightarrow \quad Ax = 0.$$

Proof Let x be a vector in E such that $q(x) = 0$. Then

$$\forall t \in \mathbb{R}, \quad y \in E, \qquad q(x + ty) = t(tq(y) + 2\,{}^t yAx) \geqslant 0$$

and ${}^t yAx = 0$ for all y. Thus, $Ax = 0$. ∎

Proof of Theorem 1 Using the previous lemma, we can choose for E_0 the linear subspace of E defined by

$$E_0 = \{x \in E : q(x) = 0\} = \{x \in E : Ax = 0\}.$$

Let F be a linear subspace of E such that

$$E_0 + F = E, \qquad E_0 \cap F = \{0\}.$$

The construction of E_0 and F guarantees that Conditions (1)–(3) are satisfied. Furthermore, the lemma implies that

$$\forall x_0 \in E_0, \quad y \in F, \qquad \varphi(x_0, y) = {}^t x_0 A y = 0. \quad ∎$$

Theorem 2 *Let E be a linear space of dimension n and $q, q_1, \ldots, q_k$ be $k + 1$ nonnegative definite quadratic forms in E having ranks $r, r_1, \ldots, r_k$ respectively. If*

$$\forall x \in E, \qquad q(x) = q_1(x) + q_2(x) + \cdots + q_k(x), \quad r = r_1 + r_2 + \cdots + r_k,$$

then there exist $k + 1$ linear subspaces of E, designated by $F, F_1, \ldots, F_k$, such that

(1) *$F, F_1, \ldots, F_k$ are supports of $q, q_1, \ldots, q_k$, respectively;*
(2) *for every $i \neq j$, F_i and F_j are orthogonal with respect to the quadratic form q;*
(3) $F = F_1 + \cdots + F_k$.

Proof We prove this theorem by induction on k. The theorem is true for $k = 1$. If it is true for $k < s$, proving it for $k = s$ is equivalent to proving it for $k = 2$. Let q, q_1, q_2 be three quadratic forms, satisfying the assumptions of the theorem when $k = 2$. Let E_0, E_1, E_2 be their respective kernels and $\varphi, \varphi_1, \varphi_2$ be the respective bilinear forms associated with these quadratic forms. Let x be a vector in $E_1 \cap E_2$. We then have

$$q(x) = q_1(x) + q_2(x),$$

whence

$$q_1(x) = q_2(x) = 0 \Rightarrow q(x) = 0.$$

Conversely, let x be a vector in E_0. Since q_1 and q_2 are nonnegative definite, we have

$$q(x) = 0 \Rightarrow q_1(x) = q_2(x) = 0$$

and then $E_1 \cap E_2 = E_0$. Now we can assume that $E_0 = \{0\}$, i.e., $r = n$. If this were not so, we could restrict ourselves to the orthogonal supplement [13, p. 121] (with respect to q) of E_0. When $r = n$, E_1 and E_2 satisfy

$$E_1 \cap E_2 = \{0\}, \qquad \dim E_1 + \dim E_2 = n$$

and

$$\forall x_1 \in E_1, \quad \forall x_2 \in E_2, \qquad \varphi(x_1, x_2) = \varphi_1(x_1, x_2) + \varphi_2(x_1, x_2) = 0.$$

Then we can choose E_1 as the support of F_2 and E_2 as the support of F_1 since for every nonzero x in E_1 we have

$$q_2(x) = q(x) > 0$$

because q is positive (rank$(q) = n$). ∎

A straightforward consequence of the previous theorem is:

Theorem 3 (Cochran) *Let $q_1, \ldots, q_k$ be k nonnegative quadratic forms on $\mathbb{R}^n$, with ranks less than or equal to $r_1, \ldots, r_k$. If*

$$q_1 + q_2 + \cdots + q_k = \mathbf{1}_{\mathbb{R}^n}, \qquad r_1 + r_2 + \cdots + r_k = n,$$

then there exists an orthonormal basis of $\mathbb{R}^n$, such that

$$q_i(y) = \sum_{j=n_{i-1}+1}^{n_i} y_j^2, \qquad i = 1, \ldots, k$$

where

$$n_0 = 0, \qquad n_1 = r_1, \qquad n_2 = r_1 + r_2, \ldots, n_i = r_1 + \cdots + r_i, \ldots, n_k = n.$$

2. Linear-Normal Statistical Spaces

Definition 1 We call

$$(\Omega, \mathscr{B}_\Omega;\ N(m, \sigma^2\Lambda), m \in V, \sigma \in \mathbb{R}^+),$$

a *linear-normal* statistical space, where Ω is a finite-dimensional linear space, $\mathscr{B}_\Omega$ is the σ-field of the Borel sets of Ω, V is a proper linear subspace of Ω, Λ is a given positive-definite quadratic form on the dual space Ω^* of Ω, and the inner product on Ω is defined by Λ^{-1} (The definition of Λ^{-1} is given below.)

The *dual space* of Ω is the linear space of all linear forms on Ω. We recall that a normal vector X, valued on the finite-dimensional linear space $\mathscr{X}$, has the characteristic function

$$\forall t \in \mathscr{X}^*, \qquad \Phi_X(t) = E(\exp[it(X)]) = \exp[it(m) - \tfrac{1}{2}\Lambda(t)],$$

where $\mathscr{X}^*$ is the dual space of $\mathscr{X}$, m a vector of $\mathscr{X}$, and Λ a nonnegative quadratic form on $\mathscr{X}^*$. Moreover, we have

$$E(x) = m,$$

$$E(t(x)^2) = \Lambda(t), \qquad \forall t \in \mathscr{X}^*.$$

We also recall that if a basis has been chosen for $\mathscr{X}$, the quadratic form Λ^{-1} is defined with respect to the dual basis of $\mathscr{X}^*$ by the matrix which is the inverse of the matrix defining Λ on $\mathscr{X}$. Finally, see [13; p. 118] for the definition of inner product by means of a quadratic form.

Linear-normal statistical spaces correspond to the usual linear models. For example, in problems of analysis of variance (see Section 7), Ω is the linear space of all the mappings from a given finite set E to $\mathbb{R}$ and V is the subspace generated by some given real functions on E.

To study a statistical problem which is defined on a linear-normal statistical space, it is of interest to distinguish the basic statistical problems from the computational ones. The former will be solved by a geometrical argument and the latter by an algebraic argument involving matrices after choosing a basis for Ω. Note that this latter argument depends strongly on how the subspace V is given.

We now recall some useful results. Let $\mathscr{X}$ be a finite-dimensional Hilbert space and let W be a linear subspace of $\mathscr{X}$. For every x in $\mathscr{X}$ we let x_W be the projection of x on W. We also let $W^\perp$ be the linear subspace of $\mathscr{X}$, orthogonal to W, such that $\mathscr{X} = W + W^\perp$; $W^\perp$ is the *orthogonal complement* of W [13, p. 121].

Theorem 1 *Let $A[n, m]$ and $B[p, m]$ be two given matrices and let M be a positive-definite symmetric matrix. Let V be the linear subspace of $\mathbb{R}^n$ defined by the vectors x such that*

$$x = A\theta, \qquad B\theta = 0, \qquad \theta \in \mathbb{R}^m. \tag{1}$$

We assume that V is different from $\{0\}$ and we introduce on $\mathbb{R}^n$ the inner product defined by M. Let $S = {}^{t}AMA$. Then

(a) *for every x in $\mathbb{R}^n$ the linear equations in (θ, μ)*

$$B\theta = 0, \qquad S\theta + {}^{t}B\mu = {}^{t}AMx \tag{2}$$

have at least one solution and for any such solution $A\theta = x_V$;

(b) *the matrix,*

$$H = \begin{pmatrix} S & {}^{t}B \\ B & 0 \end{pmatrix}$$

is nonsingular if and only if the rows of B are linearly independent and for every x in V there exists a unique θ which satisfies Condition (1)*;*

(c) *if H is nonsingular, let G be the matrix obtained from H^{-1} by deleting the last p rows and columns; then*

$$x_V = AG\,{}^t\!AMx, \qquad GSG = G, \qquad BG = 0, \tag{3}$$

and the matrix SG has exactly $m - p$ nonzero eigenvalues, all equal to 1;

(d) *if the matrices H and S are nonsingular, then*

$$G = Q - Q\,{}^t\!B(BQ\,{}^t\!B)^{-1}BQ, \tag{4}$$

where $Q = S^{-1}$.

Proof (a) x_V, the projection of x on V, is characterized by

$$\langle x - x_V \mid y \rangle = 0, \qquad \forall y \in V. \tag{5}$$

Let θ be such that $x_V = A\theta$, $B\theta = 0$. Then (5) is equivalent to

$$\theta' \in \mathbb{R}^m, \qquad B\theta' = 0 \;\Rightarrow\; {}^t(x - A\theta)MA\theta' = 0.$$

But this is equivalent to the existence of a vector μ in $\mathbb{R}^p$, such that

$${}^t(x - A\theta)MA = {}^t\mu B,$$

establishing (2) since $B\theta = 0$. Then, for every x, (2) has at least one solution and conversely, for every θ satisfying (2), $A\theta$ is always equal to x_V.

(b) If the matrix H is nonsingular, the rows of B are linearly independent and for every x, Eq. (2) has a unique solution. Specifically, if x is in V, this unique solution satisfies $x = A\theta$. Conversely if the rows $B_1, \ldots, B_p$ of B are linearly independent, then

$$\forall \mu[p, 1], \qquad {}^t\!B\mu = 0 \;\Rightarrow\; \mu = 0. \tag{6}$$

Now, consider the equations

$$B\theta = 0, \qquad S\theta + {}^t\!B\mu = 0. \tag{7}$$

Multiplying (7) on the left by ${}^t\theta$ leads to

$${}^t\theta S\theta = 0 = \|A\theta\|^2.$$

Thus (7) implies that $B\theta = 0$, $A\theta = 0$. By hypothesis, this implies that $\theta = 0$ and finally, by (6), that $\mu = 0$. Then Eq. (7) has the unique solution, $\theta = 0$, $\mu = 0$. This means that H is nonsingular.

(c) Write H^{-1} in the form,

$$H^{-1} = \begin{pmatrix} G & C_1 \\ {}^tC_1 & C_2 \end{pmatrix}, \tag{8}$$

so that

$$GS + C_1B = \mathbb{1}_n, \tag{9}$$

$$G\,{}^t\!B = 0, \qquad {}^tC_1S + C_2B = 0, \qquad BC_1 = \mathbb{1}_p. \tag{10}$$

However, the matrix G is symmetric. Then (10) implies that $GB = 0$. Equations (9) and (10) imply that $GSB = G$. Let x be an eigenvector of GS associated with the eigenvalue s. Now (9) implies that x is an eigenvector of C_1B associated with the eigenvalue $(1 - s)$. Therefore, using (10), we have

$$C_1Bx = (1 - s)x \quad \Rightarrow \quad BC_1Bx = (1 - s)Bx = Bx.$$

Thus, either s or Bx vanishes, i.e., s is either 0 or 1. This shows that all the nonzero eigenvalues of GS are equal to 1. The number of such unit eigenvalues is the dimension of the linear subspace $Bx = 0$. This number is $m - p$, since the rows of B are linearly independent.

(d) Here, S is nonsingular and it is easily seen (Exercise 4) that the matrix $BQ\,{}^tB$ is also nonsingular. Let C_1 and C_2 be given by

$$C_2 = -(BQ\,{}^tB)^{-1}, \qquad C_1 = -Q\,{}^tBC_2.$$

It is easily verified that the matrices G, C_1, C_2 (where G is given by (4)), satisfy Conditions (9) and (10). ∎

3. Fundamental Theorems

Let $\mathscr{X}$ be a finite-dimensional linear space, W a linear subspace of $\mathscr{X}$ and X a random vector, taking values on $\mathscr{X}$, having the nonsingular distribution $N(m, \Lambda)$. We introduce on $\mathscr{X}$ the inner product defined by Λ^{-1}. Then, it is obvious that X_W is a normal random vector whose expectation is m_W. We denote by Λ_W the covariance of X_W. Note that Theorem 2.1 can be used to compute the matrix which defines Λ_W in the usual cases.

Theorem 1 *Let $\mathscr{X}$ be a finite-dimensional linear space and X a random vector taking values on $\mathscr{X}$ having the nonsingular distribution $N(m, \sigma^2\Lambda)$. We introduce on $\mathscr{X}$ the inner product defined by Λ^{-1} and let $F_1, \ldots, F_k$ be k subspaces of $\mathscr{X}$ which are mutually orthogonal. Then*

(i) *The random vectors X_{F_i} $(i = 1, \ldots, k)$ are independent and have $N(m_{F_i}, \sigma^2\Lambda_{F_i})$ distributions, respectively.*

(ii) *The random variables $\|X_{F_i}\|^2$ $(i = 1, \ldots, k)$ are independent and have $\Gamma(\frac{1}{2}\dim F_i, \|m_{F_i}\|^2, 2\sigma^2)$ distributions, respectively.*

This theorem can be easily proven by various arguments (see Exercise 7). Using Cochran's theorem we deduce from Theorem 1 the following corollary which is useful in many applications.

Corollary *Let X be a random vector taking values in $\mathbb{R}^n$ having the non-*

singular $N(m, \sigma^2\Lambda)$ distribution, and let $q_1, \dots, q_k$ be k nonnegative definite quadratic forms on $\mathbb{R}^n$ such that

$$q_1 + \cdots + q_k = \Lambda^{-1},$$

$$\operatorname{rank}(q_i) \leqslant r_i, \qquad i = 1, \dots, k,$$

$$r_1 + \cdots + r_k = n.$$

Then the random variables $q_i(X)$ are independent and have $\Gamma(r_i/2, q_i(m), 2\sigma^2)$ distributions, respectively.

Theorem 2 *For the linear-normal statistical space defined in Section 2 we have*

(i) *$(x_V, \|x_{V^\perp}\|^2)$ is a complete and sufficient statistic;*
(ii) *x_V and $\|x_{V^\perp}\|^2$ are independent statistics and have $N(m, \sigma^2\Lambda_V)$ and $\Gamma(\dim V/2, 2\sigma^2)$ distributions, respectively.*

Proof The likelihood function is equal to

$$(2\pi\sigma^2)^{-\dim\Omega/2} \exp\{-(1/2\sigma^2)\|x - m\|^2\}.$$

But, $\|x - m\|^2 = \|x_V - m\|^2 + \|x_{V^\perp}\|^2$. Theorem 2.2.2 shows that $(x_V, \|x_{V^\perp}\|^2)$ is a sufficient statistic. We introduce $\theta_0 = -1/(2\sigma^2)$ and the components of m/σ^2 with respect to an orthonormal basis of V, as new parameters. Thus, it is easy to see that the sufficient statistic induces a standard exponential and complete statistical space. Using Theorem 1 the proof is completed. ∎

4. Testing Linear Hypotheses

We denote by $\Phi(\alpha; a, b)$ the value Φ for which

$$\int_{\Phi}^{+\infty} d\beta(a, b) = \alpha,$$

and consider the linear-normal statistical space of Definition 2.1.

Theorem 1 *Let W be a subspace of V and let T denote the subspace $W^\perp \cap V$. The test of the hypothesis $\{m_T = 0\}$ against the hypothesis $\{\|m_T\| > 0\}$, which is defined by the critical region*

$$\frac{\|x_T\|}{\|x_{V^\perp}\|^2} \geqslant K = \Phi\left(\alpha; \frac{\dim T}{2}, \frac{\dim V^\perp}{2}\right),$$

is called a test of the linear *hypothesis W. This test is free and has level of significance α.*

Proof The power function of this test is equal to

$$\int_K^\infty d\beta\left(\frac{\dim T}{2}, \frac{\dim V^\perp}{2}, \frac{\|m_T\|^2}{2\sigma^2}\right).$$

When m is in W, i.e., when m_T vanishes, this integral equals α. ∎

Remark This linear test has various and interesting properties. See, e.g., Hsu [29], Wald [64], and Lehmann [44, p. 268] for its power and invariance properties.

Definition 1 Let $W_1, \ldots, W_k$ be k different subspaces of V. Then the following strategy: "For every $j = 1, \ldots, k$, if

$$\frac{\|x_{W_j}\|^2}{\|x_{V^\perp}\|^2} \leqslant \Phi\left(\alpha; \frac{\dim W_j}{2}, \frac{N - \dim V}{2}\right),$$

then choose the hypothesis $\{m_{W_j} = 0\}$," is called a *test of the hypothesis* $\{W_j, j = 1, \ldots, k\}$.

Note that we do not assume the subspace W_j to be orthogonal. For instance, some of the W_j may be the sum of others; then the hypotheses to be tested may be dependent. For example, let V' and V'' be two subspaces of V, where

$$V'' \subset V' \subset V.$$

Consider the problem of testing the hypotheses $\{m \in V'\}$ and $\{m \in V''\}$. Use the strategy given by Definition 1, i.e.,

$$\|x_{W_1}\|^2/\|x_{V^\perp}\|^2 \leqslant \Phi(\alpha; \tfrac{1}{2}(\dim W_1), \tfrac{1}{2}(N - \dim V)) \quad \Rightarrow \quad \{m \in V'\}, \tag{1}$$

$$\|x_{W_2}\|^2/\|x_{V^\perp}\|^2 \leqslant \Phi(\alpha; \tfrac{1}{2}(\dim W_2), \tfrac{1}{2}(N - \dim V)) \quad \Rightarrow \quad \{m \in V''\}, \tag{2}$$

where

$$W_1 = (V')^\perp \cap V, \qquad W_2 = (V'')^\perp \cap V.$$

If Test (1) rejects the hypothesis $\{m \in V'\}$ and if Test (2) accepts the hypothesis $\{m \in V''\}$, we reject the hypothesis $\{m \in V'\}$. This strategy is considered better than the one in which (2) is replaced by

$$\|x_{W'}\|^2/\|x_{V^\perp}\|^2 \leqslant \Phi(\alpha, \tfrac{1}{2}(\dim W'), \tfrac{1}{2}(N - \dim V)) \quad \Rightarrow \quad \{m \in V''\}, \tag{3}$$

where

$$W' = (V'')^\perp \cap V'.$$

5. Estimation of Linear Functions

Again, we consider the linear-normal statistical space of Definition 2.1.

Theorem 1 *The statistic $x \to x_V$ is an unbiased estimate of m having minimum variance. Moreover, the function*

$$G(x, m) = \|x_V - m\|^2/\|x_{V^\perp}\|^2$$

is a pivotal function for the set estimation of m and has the $\beta(\frac{1}{2}\dim V, \frac{1}{2}\dim V^\perp)$ distribution.

Proof The statistic $x \to (x_V, \|x_{V^\perp}\|^2)$ is sufficient and complete. Then x_V, which has the $N(m, \sigma^2 \Lambda_V)$ distribution, is the unique unbiased estimator of m which depends only on the sufficient statistic. It is easy to see that G is a pivotal function (Definition 6.4.1). ∎

Remark We have seen previously (Section 3) that

$$\log \mathscr{L}(m, \sigma; x) = -\tfrac{1}{2}\dim \Omega \log(2\pi\sigma^2) - \frac{1}{2\sigma^2}\|x_{V^\perp}\|^2 - \frac{1}{2\sigma^2}\|x_V - m\|^2.$$

Thus the maximum likelihood estimators of m and σ^2 are x_V and $\|x_{V^\perp}\|^2/\dim \Omega$, respectively.

Theorem 2 *Let f be a linear mapping from V into a finite-dimensional linear space. Then $f(x_V)$ is the unbiased estimate of $f(m)$ having minimum variance. Let W be the subspace of V which is the orthogonal complement of the kernel of f and let H be the function defined by*

$$x \in \Omega, \qquad y \in f(V) \to H(x, y) = \|x_W - m(y)\|^2/\|x_{V^\perp}\|^2$$

where $m(y)$ is defined by

$$m(y) \in W, \qquad f(m(y)) = y. \tag{1}$$

Then H is a pivotal function for the set estimation of $f(m)$ and has the $\beta(\frac{1}{2}(\dim W), \frac{1}{2}(\dim V^\perp))$ distribution.

Proof Since f is a linear function, we have

$$E(f(x_V)) = f(E(x_V)) = f(m).$$

Then $f(x_V)$ is the unique unbiased estimate of $f(m)$ which depends only on the sufficient statistic. It is obvious that (1) defines, without ambiguity, the mapping $y \in f(V) \to m(y) \in W$ and we have $m(f(m)) = m_W$. The proof should now be completed by the reader. ∎

Remark We have

$$\|x_W - m(y)\|^2 = Q(f(x_V) - y)$$

where Q is the quadratic form defined on $f(V)$ by

$$Q(y) = \|m(y)\|^2 = \inf(\|m\|^2 \mid m \in W, f(m) = y).$$

The covariance of $f(x_V)$ is $\sigma^2 Q^{-1}$.

Theorem 3 *Consider the statistical space*

$$[\mathbb{R}^n, \mathscr{B}_{\mathbb{R}^n}, N(A\theta, \sigma^2\Lambda); \theta \in \{B\theta = 0\}, \sigma \in \mathbb{R}^+],$$

where $A[n, m]$, $B[p, m]$, *and* Λ *are given matrices such that*

$$\begin{cases} p < m < n, \ \Lambda \text{ is a symmetric positive-definite matrix of order } n; \\ \{A\theta = 0, B\theta = 0\} \ \Rightarrow \ \theta = 0; \\ \text{the rows of } B \text{ are linearly independent.} \end{cases}$$

Denote the matrix ${}^{t}A\Lambda^{-1}A$ *by* S *and let* G *be the matrix obtained from*

$$\begin{pmatrix} S & {}^{t}B \\ B & 0 \end{pmatrix}^{-1}$$

by deleting the last p *rows and columns. Then*

(i) *The statistic*

$$\hat{\theta} = G\,{}^{t}A\Lambda^{-1}x$$

is the maximum likelihood and unbiased estimate of θ, *having minimum variance, equal to* $\sigma^2 G$.

(ii) *The statistic,*

$$\hat{\sigma}^2 = \frac{1}{n + p - m}\,{}^{t}(x - A\hat{\theta})\Lambda^{-1}(x - A\hat{\theta})$$

is the unbiased estimator of σ^2 *having minimum variance. It has the distribution*

$$\Gamma\left(\frac{n + p - m}{2}, \frac{2\sigma^2}{n + p - m}\right).$$

(iii) *The function*

$$\frac{1}{(n + p - m)\hat{\sigma}^2}\,{}^{t}(\theta - \hat{\theta})S(\theta - \hat{\theta})$$

is a pivotal function for the set estimation of θ and has the distribution

$$\beta\left(\frac{m-p}{2}, \frac{n+p-m}{2}\right).$$

(iv) *$D\hat{\theta}$ is the maximum likelihood unbiased estimate of $D\theta$ having minimum variance, where $D[q, m]$ is a given matrix. Moreover, if the rows of the matrix $\binom{D}{B}$ are linearly independent, we denote by $K[m, q]$ the matrix obtained from*

$$\begin{pmatrix} S & {}^{t}B & {}^{t}D \\ B & 0 & 0 \\ D & 0 & 0 \end{pmatrix}^{-1}$$

by keeping only the first m rows of the q last columns. Then the function

$$\frac{1}{(n+p-m)\hat{\sigma}^2}\,{}^{t}(D\hat{\theta} - D\theta)\,{}^{t}KSK(D\hat{\theta} - D\theta)$$

is a pivotal function for the set estimation of $D\theta$ and has the distribution

$$\beta\left(\frac{q}{2}, \frac{n+p-m}{2}\right).$$

Proof Apply Theorem 2.1 to obtain the formulas corresponding to the results of Theorems 1 and 2. Here

$$\dim V = m - p$$

and the kernel V' of the linear mapping $\{x \in V \to D\theta\}$ is defined by

$$x = A\theta, \qquad B\theta = 0, \qquad D\theta = 0.$$

Then apply Theorem 2.1 with $\binom{B}{D}$ instead of B and obtain

$$x_{V'} = AG'\,{}^{t}A\Lambda^{-1}x$$

where G' is defined by

$$\begin{pmatrix} S & {}^{t}B & {}^{t}D \\ B & 0 & 0 \\ D & 0 & 0 \end{pmatrix}^{-1} = \begin{pmatrix} G' & C_1' & K \\ {}^{t}C_1' & J & L \\ {}^{t}K & {}^{t}L & N \end{pmatrix} \tag{2}$$

since Theorem 2.1 shows that this matrix is nonsingular. Now let

$$W = (V')^{\perp} \cap V.$$

Then

$$x_W = x_V - x_{V'} = A(G - G')\,{}^{t}A\Lambda^{-1}x.$$

But W is a subspace of V and then

$$x_W = A(G - G')\,{}^tA\Lambda^{-1}x_V = A(G - G')S\hat{\theta}.$$

Using concepts introduced in the proof of Theorem 2.1 (c) it can be seen that there exists a matrix C_1 such that $GS + C_1B = \mathbb{1}_m$. Now, using (2), we obtain $G'S + C_1'B + KD = \mathbb{1}_m$ and then

$$(G - G')S + (C_1 - C_1')B = KD.$$

Multiplication on the right by $\hat{\theta}$ yields

$$(G - G')S\hat{\theta} = KD\hat{\theta},$$

since $B\hat{\theta} = 0$. Finally, we have

$$\|x_W\|^2 = \|AKD\hat{\theta}\|^2 = {}^t(D\hat{\theta})\,{}^tK\,{}^tA\Lambda^{-1}AKD\hat{\theta} = {}^t\hat{\theta}\,{}^tD\,{}^tKSKD\hat{\theta}.$$

The proof is completed by noting that the same computation is valid if we use θ instead of $\hat{\theta}$. ∎

Remark When applying the previous formulas for constructing confidence regions of θ or of $D\theta$, do not forget the constraint $B\theta = 0$.

6. Fundamental Lemmas for Analysis of Variance

In this section we use the following notations: k is an integer greater than unity; K is the set of the integers $1, \ldots, k$; $v_1, \ldots, v_k$ are k given positive integers. We denote by E the finite set

$$E = \prod_{j=1}^{k} \{1, \ldots, v_j\}$$

and by $e = (e_1, \ldots, e_k)$ an element of E. Then, for every $j = 1, \ldots, k$ the component e_j can take on the integral values $1, \ldots, v_j$. For every finite set F we denote by $v(F)$ the number of elements of F. $\mathscr{X}$ is the finite-dimensional linear space defined by all mappings from E to $\mathbb{R}$. We introduce on $\mathscr{X}$ the inner product,

$$m, m' \in \mathscr{X}, \qquad \langle m | m' \rangle = \sum_{e \in E} m(e)m'(e).$$

The following results are fundamental for *both* Model I and analysis of variance, Model II.

Definition 1 To every function m belonging to $\mathscr{X}$, we associate the functions,

$$\forall J \subset K, \qquad S_J(e) = \begin{cases} m(e) & \text{if } J = \varnothing, \\ \dfrac{1}{\prod_{j \in J} v_j} \sum_{e_j : j \in J} m(e) & \text{if } J \neq \varnothing, \end{cases}$$

$$\forall I \subset K, \qquad m_I = \sum_{J \subset \complement I} (-1)^{v(J - \complement I)} S_J.$$

The formula $\sum_{e_j : j \in J} m(e)$ means that the summation is taken over all possible values of the components e_j whose indices are in J. For example,

$$S_{\{1,2\}}(e) = \frac{1}{v_1 v_2} \sum_{e_1=1}^{v_1} \sum_{e_2=1}^{v_2} m(e_1, \ldots, e_k).$$

Remark The function $S_J(e)$ depends only on the components e_j of e whose indices are not in J; then m_I depends only on the components e_j of e whose indices are in I. It can be verified that

$$\forall I' \neq I \subset K, \qquad (m_I)_{I'} = 0,$$
$$\forall I \subset K, \qquad (m_I)_I = m_I.$$

Lemma 1 *The following properties are true:*

(i) $m_\varnothing$ *is a constant function;*

(ii) $m(e) = \sum_{I \subset K} m_I(e);$ (1)

(iii) $\sum_{e_j=1}^{v_j} m_I(e) = 0, \qquad \forall j \in I \neq \varnothing.$ (2)

Proof The functions $S_J(e)$ are such that

$$\forall j \in K, \qquad \sum_{e_j=1}^{v_j} S_J = \begin{cases} v_j S_J, & \text{if } j \in J \\ v_j S_{J+\{j\}}, & \text{if } j \notin J. \end{cases} \tag{3}$$

We have $m_\varnothing = S_K$ and for every nonempty I

$$\forall j \in I, \qquad \sum_{e_j=1}^{v_j} m_I = \sum_{J \supset \complement I} (-1)^{v(J - \complement I)} \sum_{e_j=1}^{v_j} S_J. \tag{4}$$

Treating the two cases, $j \in J$ and $j \notin J$ in the right-most summation of (4) by using (3) we obtain

$$\sum_{e_j=1}^{v_j} m_j = \sum_{J \supset \complement I + \{j\}} (-1)^{v(J - \complement I)} v_j S_J + \sum_{j \notin J \supset \complement I} (-1)^{v(J - \complement I)} v_j S_{J+\{j\}}. \tag{5}$$

Letting $J' = J + \{j\}$ in the right-most summation of (5) we obtain

$$\sum_{J \supset \complement I + \{j\}} (-1)^{\nu(J - \complement I)} v_j S_J - \sum_{J' \supset \complement I + \{j\}} (-1)^{\nu(J' - \complement I)} v_j S_{J'} = 0. \tag{6}$$

Then, using I' for $\complement I$ yields

$$\sum_I m_I = \sum_{J \subset K} S_J \Bigg(\sum_{I' \subset J} (-1)^{\nu(J - I')} \Bigg). \tag{7}$$

But we know that for every nonempty finite set X,

$$\sum_{x \in \mathscr{P}(X)} (-1)^{\nu(x)} = 0.$$

Then, all terms in the right-hand side of (7) vanish except $S_\varnothing$ which is equal to m. ∎

Definition 2 For every subset I of K we designate by $\mathscr{X}_I$ the linear subspace of $\mathscr{X}$ defined by the function m, where

$$m_J = 0, \qquad \forall J \neq I.$$

Lemma 2 *When I runs through all the subsets of K which are different from K, the subspaces $\mathscr{X}_I$ are orthogonal and generate $\mathscr{X}$. Moreover, for every function m in $\mathscr{X}$, m_I is the projection of m on $\mathscr{X}_I$.*

Proof Let I and I' be two distinct subsets of K. Without loss of generality we can assume that $I - (I' \cap I)$ is nonempty and then let j be an index belonging to this set. For every m and m' belonging to $\mathscr{X}_I$ and $\mathscr{X}_{I'}$ respectively, we have

$$m = m_I, \qquad m' = m'_{I'}.$$

Then using (2), we obtain

$$\langle m | m' \rangle = \langle m_I | m'_{I'} \rangle = \sum_e \Bigg(m'_{I'}(e) \sum_{e_j} m_I(e) \Bigg) = 0,$$

i.e., $\mathscr{X}_I$ and $\mathscr{X}_{I'}$ are orthogonal. The proof is easily completed by using (1). ∎

Lemma 3 *For every I of K ($I \neq K$) let $\mathscr{F}_I$ be the subspace of $\mathscr{X}$ defined by the functions which depend only on the components e_j, $j \in I$. Let $\mathscr{F}_I^0$ be the subspace of $\mathscr{F}_I$ generated by all the subspaces $\mathscr{F}_{I'}$ where*

$$I' \subset I, \qquad I' \neq I. \tag{8}$$

Then we have

$$\mathscr{X}_I = \mathscr{F}_I \cap (\mathscr{F}_I^0)^\perp.$$

Proof Let m be a function in $\mathscr{X}_I$, let m' be a function in $\mathscr{F}_{I'}$ where the subset I' satisfies (8), and let j be an element of $I - I'$. The same argument as used in Lemma 2 shows that

$$\langle m | m' \rangle = \sum_e m(e) m'(e) = \sum_e \left(m'(e) \sum_{e_j} m(e) \right) = 0$$

and then m is orthogonal to $\mathscr{F}_I^0$. Conversely, let m be a function in $\mathscr{F}_I$ which is orthogonal to every $\mathscr{F}_{I'}$ where I' satisfies (8). Since $\mathscr{X}_{I'}$ is included in $\mathscr{F}_{I'}$, this function m is orthogonal to every $\mathscr{X}_{I'}$ where I' satisfies (8). Using the same argument, we see that m is also orthogonal to every $\mathscr{X}_{I'}$ in which $I' - I' \cap I \neq \varnothing$. Thus, m is in $\mathscr{X}_I$. ∎

Remark Lemma 3 shows that (1) defines a decomposition of m, where each function m_I is a function which depends only on components e_i, $i \in I$, and cannot be written as a sum of functions which do not depend on all these components.

7. Methodology of Analysis of Variance (Model I) in Experimental Design

We now discuss some statistical problems which can be treated by analysis of variance (Model I). Let X be a random variable having the $N(m, \sigma^2)$ distribution. We assume that we can observe X under different circumstances and that

(a) The mean m of X depends in an unknown way on a finite number of *factors*, $F_1, \ldots, F_k$, each of which can take on a finite number of values called *levels*. Let F_j^i be the ith level of the jth factor for every $j = 1, \ldots, k$; the index i takes on values $1, \ldots, v_j$ where the integers v_j are given and are greater than one.

(b) The variance σ^2 does not depend on the factors, and is unknown.

(c) The correlation coefficient between any two observations on X is known.

The set of all combinations of the factor levels for which we have one or more observations on X is called an *experimental design*.

We now consider the problem of testing some suitable hypotheses on the function

$$\{F_j^i; j = 1, \ldots, k, i = 1, \ldots, v_j\} \rightarrow m$$

and find conditions under which we can apply the general theory developed in previous sections.

For example, the following problems can be solved by means of analysis of variance methods:

(a) rainfall as a function of the factors, country and season;
(b) agricultural yield as a function of the factors, time, weather and soil conditions;
(c) intellectual quotient of an individual, as a function of the factors, education, social level, creativity, etc.

Without loss of generality we formalize the concept by:

Definition 1 Let $e = (e_1, \ldots, e_k)$ $(k > 1)$ be a point of the set

$$E = \prod_{j=1}^{k} \{1, 2, \ldots, v_j\},$$

which represents all possible combinations of the levels of the factors. A subset E^* of E, such that for every $j \in K = \{1, \ldots, k\}$ and for every integer $i \in \{1, \ldots, v_j\}$ there exists an element e of E^* such that $e_j = i$, is called an *experimental design.* When $E^* = E$, the design is called a *factorial design.*

Now, we consider the mean m which is an unknown function of the factors, i.e., an unknown element of the linear space $\mathscr{X}$ of all real functions on E. Using the results of the previous section we can decompose m in the form

$$m = \sum_{I \subset K} m_I.$$

In an analysis of variance problem one always assumes that some of the m_I are equal to zero, i.e.,

$$m(e) = \sum_{I \in \mathscr{K}} m_I(e), \qquad \forall e \in E, \tag{1}$$

where $\mathscr{K}$ is a given family of subsets of K. We wish to test hypotheses,

$$H_I = \{m_I = 0\}, \qquad I \in \mathscr{K}_0, \tag{2}$$

where $\mathscr{K}_0$ is a given subfamily of $\mathscr{K}$. In general, the problem is a multiple decision one.

In order to understand the general problem of analysis of variance, one must distinguish carefully between the function m as a parameter and the mean of the observations. The function m is defined on E and the mean of the observations is a function on E^*. It is fundamental to understand that the statistician wants to have information on m as a function on E, but he can only observe X on E^*. The experimental design E^* is usually chosen by the statistician before data are collected. In some problems the set E^* cannot be given in advance, e.g., in cases where some observations have been lost.

Now, consider again the hypotheses which are treated in analysis of variance. Both assumptions and hypotheses to be tested are expressed by the vanishing of some m_I, i.e., they are linear hypotheses on m. The following hypotheses are tested frequently.

1. The function m does not depend on the factor F_j;
2. The function m depends only on the factors F_j, where j belongs to a given subset I_0;
3. There is *no interaction* between two particular factors, F_j and $F_{j'}$, i.e., F_j and $F_{j'}$ have an *additive* effect on m;
4. There is *no total interaction* among the factors F_j, where j belongs to a given subset I_0;
5. There is *no interaction* between the family of factors F_j, $j \in I_0$ and the family of factors F_j, $j \in I'_0$, where I_0 and I'_0 are two given disjoint subsets of K;

The reader can verify easily that these hypotheses are described, mathematically, by

1′. $I \supset \{j\} \quad \Rightarrow \quad m_I = 0,$
2′. $I \cap \complement I_0 \neq \varnothing \quad \Rightarrow \quad m_I = 0,$
3′. $I \supset \{j, j'\} \quad \Rightarrow \quad m_I = 0,$
4′. $I \supset I_0 \quad \Rightarrow \quad m_I = 0,$
5′. $\{I \cap I_0 \neq \varnothing, I \cap I'_0 \neq \varnothing\} \quad \Rightarrow \quad m_I = 0,$

and should note that 5 is not the union of hypothesis of Type 3.

We restate the previous basic problem; the hypotheses which have practical interest express m as a function on E, but the hypotheses which can be tested by statistical theory define m as a function on E^*. Thus it is obvious that some conditions on E^*, $\mathscr{K}$, $\mathscr{K}_0$, must be satisfied to make the problem consistent.† Let $\mathscr{X}_{\mathscr{K}}$ be the subspace of $\mathscr{X}$ defined by functions m satisfying (1) and let $\mathscr{X}^*$ be the linear space of all real functions on E^*. We designate by R the mapping from $\mathscr{X}$ to $\mathscr{X}^*$, defined by

$$m \in \mathscr{X} \to m^* \in \mathscr{X}^*,$$

where

$$\forall e \in E^*, \qquad m^*(e) = m(e),$$

this mapping is called the *restriction* on E^*. Let $\mathscr{X}^*_{\mathscr{K}} = R(\mathscr{X}_{\mathscr{K}})$ and let $W_I(I \in \mathscr{K}_0)$ be the subspace of $\mathscr{X}_{\mathscr{K}}$ defined by adding the condition

$$m_I = 0.$$

† If E^*, $\mathscr{X}$, $\mathscr{X}_0$ are such that the problem is consistent, we say that $(E^*, \mathscr{X}, \mathscr{X}_0)$ is consistent. Note that $\mathscr{X}^*$ in this section is *not* the dual space of $\mathscr{X}$ as in Section 2. (Ed.)

Then, the *conditions of consistency* are

(i) For every $m^* \in \mathscr{X}^*_{\mathscr{K}}$ there exists a unique $m \in \mathscr{X}_{\mathscr{K}}$ such that $R(m) = m^*$.
(ii) $\{m \in \mathscr{X}_{\mathscr{K}}, R(m) \in R(W_I)\} \quad \Rightarrow \quad m_I = 0$.
(iii) The total number of observations is greater than $\dim \mathscr{X}_{\mathscr{K}}$.

Note that the last condition is necessary in order to construct linear tests. Instead of (i) and (ii) we often use algebraic conditions involving $\mathscr{K}$, $\mathscr{K}_0$. These conditions are linear with respect to m and in a given problem they are verified by showing that a particular determinant does not vanish.

Now we can define the statistical spaces which correspond to some typical problems of analysis of variance (Model I). For brevity we shall frequently associate $\mathscr{X}^*$ with its dual space, the linear space of linear forms on $\mathscr{X}^*$.

Problem 1 For every $e \in E^*$, only one observation has been taken. All these observations are independent. Then the corresponding statistical space is

$$(\mathscr{X}^*, \mathscr{B}_{\mathscr{X}^*}; N(m^*, \sigma^2 \mathbf{1}_{\mathscr{X}^*}), \sigma \in \mathbb{R}^+, m^* \in \mathscr{X}^*_{\mathscr{K}}),$$

where $\mathbf{1}_{\mathscr{X}^*}$ is the unit quadratic form on the dual space of $\mathscr{X}^*$.

Problem 2 For every $e \in E^*$, $n(e)$ observations have been taken. All these observations, when e runs through E^*, are independent. We assume that

$$q = \sum_{e \in E^*} (n(e) - 1) > 0.$$

Let $\{x_i(e); i = 1, \ldots, n(e)\}$ be the observations corresponding to the element e of E^*, which define a sample of size $n(e)$. Thus the following statistic is sufficient

$$x(e) = \frac{1}{n(e)} \sum_{i=1}^{n(e)} x_i(e), \qquad y(e) = \sum_{i=1}^{n(e)} [x_i(e) - x(e)]^2.$$

When e runs through E^*, the statistics $x(e)$ and $y(e)$ are independent. Furthermore, $x(e)$ has the $N(m(e), \sigma^2/n(e))$ distribution and $y(e)$ has the $\Gamma((n(e) - 1)/2, 2\sigma^2)$ distribution. In addition it is clear that the set of statistics $\{y(e), e \in E^*\}$ admits the following sufficient statistic

$$S = \sum_{e \in E^*} y(e),$$

which has the $\Gamma(q/2, 2\sigma^2)$ distribution. Now let Λ be the quadratic form on $\mathscr{X}^*$ defined by

$$\forall t \in \mathscr{X}^*, \qquad \Lambda(t) = \sum_{e \in E^*} \frac{t^2(e)}{n(e)}.$$

It is easy to see that the statistical space induced by $\{x(e), e \in E^*; S\}$ is

$$[X^*, \mathscr{B}_{\mathscr{X}^*}; N(m^*, \sigma^2 \Lambda), \sigma \in \mathbb{R}^+, m^* \in \mathscr{X}^*_{\mathscr{K}}] \times [\mathbb{R}^+, \mathscr{B}_{\mathbb{R}^+}; \Gamma(q/2, 2\sigma^2), \sigma \in \mathbb{R}^+].$$

Problem 3 For every $e \in E^*$, only one observation $x(e)$ has been taken and the correlation coefficient between $x(e)$ and $x(e')$ is a known function $\rho(e, e')$ on $E^* \times E^*$.

Let Λ' be the quadratic form on $\mathscr{X}^*$ defined by

$$\forall t \in \mathscr{X}^*, \qquad \Lambda'(t) = \sum_{e, e' \in E^*} \rho(e, e')t(e)t(e').$$

This problem is the same as Problem 1 where $\mathbf{1}_{\mathscr{X}^*}$ is replaced by Λ'.

For all three problems, note that if we have only one observation for each $e \in E^*$, the statistical space is

$$\Sigma = [\mathscr{X}^*, \mathscr{B}_{\mathscr{X}^*}; N(m^*, \sigma^2 \Lambda_0), \sigma \in \mathbb{R}^+, m^* \in \mathscr{X}^*_{\mathscr{K}}],$$

where Λ_0 is a given quadratic form on $\mathscr{X}^*$. If we have more observations (E^* remaining the same) by using sufficient statistics, we obtain the same statistical space (with a different Λ_0). Moreover we have independently obtained a statistic S which is used to estimate σ^2. Finally, it is the statistical space Σ which is basic in analysis of variance problems. Since we have seen that the hypotheses to be tested are linear, the methods of Section 4 (Definitions 1 and 2) can be applied. Note that the inner products on $\mathscr{X}$ (parameter space) and on $\mathscr{X}^*$ (observation space) are in general not the same.

It is difficult to be explicit about the relationship of consistency of $(E^*, \mathscr{K}, \mathscr{K}_0)$. For instance, for given $(\mathscr{K}, \mathscr{K}_0)$ there exists a minimal (in an algebraic sense) experimental design such that $(E^*, \mathscr{K}, \mathscr{K}_0)$ is consistent. But the practical computation of these minimal designs is a difficult combinatorial problem. Conversely, if E^* is different from E and if only one observation on each $e \in E^*$ has been taken, $\mathscr{K}$ cannot be taken as the family of all subsets of K, i.e., we must assume some prior restriction on the function m.

In the following section we treat some classical and simple cases of analysis of variance. In these cases $R(W_I)$, where $I \in \mathscr{K}_0$, are orthogonal subspaces of $\mathscr{X}^*$ with respect to the inner product defined by Λ^{-1}. Then, using Cochran's theorem or the results of the previous section makes the computations much simpler.

8. Analysis of Variance for Some Standard Experimental Designs

First, we consider the case of a *factorial design* (*of experiment*) *with two factors*. Here we let $r = v_1$, $s = v_2$ and assume that we have n observations on each of the rs elements of E, and that all of these observations are independent.

If $n = 1$, the statistical space which describes this statistical experiment is

$$(\mathbb{R}^{rs}, \mathscr{B}_{\mathbb{R}^{rs}}; N(m, \sigma^2 \mathbb{1}_{rs}), m \in \mathbb{R}^{rs}, \sigma \in \mathbb{R}^+).$$

If $n > 1$, let x_{ij}^l ($i = 1, \ldots, r, j = 1, \ldots, s, l = 1, \ldots, n$) be the given observations. Then the following set of statistics is sufficient:

$$x_{ij} = \frac{1}{n} \sum_{l=1}^{n} x_{ij}^l$$

$$s_0 = \sum_{i=1}^{r} \sum_{j=1}^{s} \left(\sum_{l=1}^{n} (x_{ij}^l - x_{ij})^2 \right).$$

This last case differs from the case: $n = 1$ only by the use of s_0, which gives an estimate of σ^2.

If m is a vector of $\mathbb{R}^{rs}$ given in the form

$$\{m_{ij}; i = 1, \ldots, r, j = 1, \ldots, s\},$$

we can apply the results of Section 6. We have

$$m_{ij} = m_0 + m_i + m'_j + m'_{ij}, \qquad i = 1, \ldots, r, j = 1, \ldots, s,$$

where m_0 is a constant,

$$m_0 = \frac{1}{rs} \sum_{ij} m_{ij};$$

m_i is either zero or a nonconstant function of i,

$$m_i = \frac{1}{s} \sum_j (m_{ij} - m_0);$$

m'_j is either zero or a nonconstant function of j,

$$m'_j = \frac{1}{r} \sum_i (m_{ij} - m_0);$$

and m'_{ij} is either zero or a function which cannot be written as the sum of a function of i and a function of j.

Lemma 6.2 implies that the following subspaces of $\mathbb{R}^{rs}$ are orthogonal and generate $\mathbb{R}^{rs}$:

$$W_1 = \{m : m_0 = m_i = m'_j = 0\},$$

$$W_2 = \{m : m_0 = m'_j = m'_{ij} = 0\},$$

$$W_3 = \{m : m_0 = m_i = m'_{ij} = 0\},$$

$$W_4 = \{m : m_i = m'_j = m'_{ij} = 0\}.$$

Then it is easy to see that

$$\dim W_1 = (r-1)(s-1), \qquad \dim W_2 = r - 1,$$

$$\dim W_3 = s - 1, \qquad \dim W_4 = 1.$$

Furthermore, the quadratic forms, corresponding to the norms of the projections on each of the four previous subspaces, are

$$q_1(m) = \|m'_{ij}\|^2 = \sum_{ij} (m'_{ij})^2 = \sum_{ij} \left(m_{ij} - \frac{1}{s} \sum_j m_{ij} - \frac{1}{r} \sum_i m_{ij} + \frac{1}{rs} \sum_{ij} m_{ij} \right)^2,$$

$$q_2(m) = \|m_i\|^2 = \sum_{ij} m_i^2 = \frac{1}{s} \sum_i \left(\sum_j (m_{ij} - \frac{1}{rs} \sum_{ij} m_{ij}) \right)^2,$$

$$q_3(m) = \|m'_j\|^2 = \sum_{ij} m_j'^2 = \frac{1}{r} \sum_j \left(\sum_i \left(m_{ij} - \frac{1}{rs} \sum_{ij} m_{ij} \right) \right)^2,$$

$$q_4(m) = \|m_0\|^2 = \frac{1}{rs} \left(\sum_{ij} m_{ij} \right)^2.$$

Now let x be the set of observations $\{x_{ij} : i = 1, \ldots, r, j = 1, \ldots, s\}$, as defined previously. Then the following statistics

$$S_0, S_1 = q_1(X), \quad S_2 = q_2(X), \quad S_3 = q_3(X), \quad S_4 = q_4(X)$$

are independent and distributed respectively as

$$\Gamma\left(\frac{rs(n-1)}{2}, \frac{2\sigma^2}{n} \right), \Gamma\left(\frac{(r-1)(s-1)}{2}, q_1(m), \frac{2\sigma^2}{n} \right),$$

$$\Gamma\left(\frac{r-1}{2}, q_2(m), \frac{2\sigma^2}{n} \right), \Gamma\left(\frac{s-1}{2}, q_3(m), \frac{2\sigma^2}{n} \right), \Gamma\left(\frac{1}{2}, q_4(m), \frac{2\sigma^2}{n} \right).$$

Note that, in this case, $\mathscr{X}^* = \mathscr{X}$ (since $E = E^*$) and that the same "formulas" hold for m as for x since the inner products on m and x are the "same" because the observations are independent.

The statistical hypotheses considered in practice are:

H_1: The two factors are independent, i.e., $m'_{ij} = 0$ or the linear hypothesis defined by the subspace $W_1^\perp = W_2 + W_3 + W_4$;

H_2: The first factor has no influence, i.e., $m_i = m'_{ij} = 0$ or the linear hypothesis defined by the subspace $W_3 + W_4$;

H_3: The second factor has no influence, i.e., $m'_j = m'_{ij} = 0$ or the linear hypothesis defined by the subspace $W_2 + W_4$;

H_4: The two factors have no influence, i.e., $m'_i = m'_j = m'_{ij} = 0$ or the linear hypothesis defined by the subspace W_4.

Various linear tests which can be formulated are based on the following rule. Let $\mathscr{I}_0$ be a subset of $\{1, 2, 3, 4\}$ (if $n = 1$, $\mathscr{I}_0$ cannot be empty) and let $\mathscr{I}_1$ be a nonempty subset of $\complement \mathscr{I}_0$. If we assume, without testing, that

$$\{q_i(m) = 0, \qquad \forall i \in \mathscr{I}_0\},$$

then we decide that

$$\{q_i(m) = 0, \forall i \in \mathscr{I}_1\},$$

when we observe that

$$\sum_{i \in \mathscr{I}_1} s_i \Big/ \Big(s_0 + \sum_{i \in \mathscr{I}_0} s_i \Big).$$

is less than a suitable constant.

Applications of this rule, yielding some standard tests, will now be given:

(a) If n is greater than 1, we can test H_1 by using the critical region,

$$\frac{s_1}{s_0} \geqslant \Phi\left(\alpha; \frac{(r-1)(s-1)}{2}, \frac{rs(n-1)}{2} \right).$$

(b) If $n = 1$, and if we assume that H_1 is true, we can test H_2 by using the critical region

$$\frac{s_2}{s_1} \geqslant \Phi\left(\alpha; \frac{r-1}{2}, \frac{(r-1)(s-1)}{2} \right).$$

(c) If n is greater than 1, we can test H_2 by using the critical region

$$\frac{s_2}{s_0} \geqslant \Phi\left(\alpha; \frac{r-1}{2}, \frac{rs(n-1)}{2} \right).$$

(d) If n is greater than 1 and if we assume that H_1 is true, we can test H_2 by using the critical region

$$\frac{s_2}{s_0 + s_1} \geqslant \Phi\left(\alpha; \frac{r-1}{2}, \frac{(r-1)(s-1) + rs(n-1)}{2} \right).$$

(e) If $n = 1$ and if we assume that H_1 is true, we can test H_4 by using the critical region

$$\frac{s_2 + s_3}{s_1} \geqslant \Phi\left(\alpha; \frac{r+s}{2} - 1, \frac{(r-1)(s-1)}{2} \right).$$

(f) If n is greater than 1, we can test H_4 by using the critical regions,

$$\frac{s_1 + s_2 + s_3}{s_0} \geqslant \Phi\left(\alpha; \frac{r+s}{2} - 1, \frac{rs(n-1)}{2} \right).$$

(g) If n is greater than 1, and if we assume that H_1 is true, we can test H_4 by using the critical region,

$$\frac{s_2 + s_3}{s_0 + s_1} \geqslant \Phi\left(\alpha; \frac{r+s}{2} - 1, \frac{rs(n-1) + (r-1)(s-1)}{2}\right).$$

(h) If $n = 1$, and if we assume that H_1 is true we can test H_2 and H_3 by the following strategy:

$$\frac{s_2}{s_1} \leqslant \Phi\left(\alpha; \frac{r-1}{2}, \frac{(r-1)(s-1)}{2}\right) \Rightarrow H_2,$$

$$\frac{s_3}{s_1} \leqslant \Phi\left(\alpha; \frac{s-1}{2}, \frac{(r-1)(s-1)}{2}\right) \Rightarrow H_3.$$

Now we consider the case of analysis of variance of an experimental design called an Euler square.

Definition 1 An *Euler square of order* p and degree $k(k > 1)$ is a subset E^* of the set $E = \{1, \ldots, p\}^k$ (i.e., the set of the sequences $e = (e_1, \ldots, e_k)$ where every integer e_j runs independently through the integers from 1 to p), such that

(a) for all i, j $(i = 1, \ldots, k, j = 1, \ldots, p)$, E^* contains exactly p elements such that $e_i = j$.

(b) for all $i \neq i', j, j'$ $(i, i' = 1, \ldots, k; j, j' = 1, \ldots, p)$ there exists a unique element of E^* such that $e_i = j$, $e_{i'} = j'$.

When $k = 2$, $E^* = E$ and we have a standard factorial design; when $k = 3$ the design is called a *Latin square*; when $k = 4$ the design is called a *Graeco-Latin square*.

The basic results and the bibliography on Euler squares are given in Barra [4]. For example one can show that

(i) $\nu(E^*) = p^2$;
(ii) $k \leqslant p + 1$;
(iii) If there exists an Euler square of order p and degree $k > 2$, then there exists an Euler square of order p, which has a degree less than k.

Theorem 1 *Consider an analysis of variance problem in which the following are assumed:*

(i) *There are k factors, $k > 2$.*
(ii) *Each factor can have p different levels, $p > 2$.*
(iii) *All factors are independent with respect to m, i.e., all interactions are zero.*

(iv) *There exists an Euler square of order p and degree k.*

(v) *The same number of observations are taken on each element of the design and these observations are independent.*

Then, using the experimental design defined by E^*, the k hypotheses corresponding to the effect of each of the factors, are orthogonal in $\mathscr{X}^*_{\mathscr{X}}$.

To prove this theorem we shall need the following lemma, which uses notation introduced in Section 6. However E is now the set given in Definition 1.

Lemma 1 *Let m be a real function on E such that*

$$\forall e \in E, \qquad m(e) = m_\varnothing + \sum_{i=1}^{k} m_{\{i\}}(e_i) \tag{1}$$

and let E^* be an Euler square of E. Then we have

$$m_\varnothing = \frac{1}{p^2} \sum_{e \in E^*} m(e) \tag{2}$$

$$pm_{\{i\}}(e_i) = \sum_{e \in E_i^*(e_i)} (m(e) - m_\varnothing) = \sum_{e \in E_i^*(e_i)} m(e) - pm_\varnothing, \qquad i = 1, \ldots, k, \tag{3}$$

where $E_i^*(e_i)$ is the subset of E^*, in which the ith *component is e_i.*

Proof Using (a) of Definition 1 we have

$$\sum_{e \in E^*} m_{\{i\}}(e_i) = p \sum_{e_i = 1}^{p} m_{\{i\}}(e_i).$$

But these quantities are zero (Lemma 6.1). Now summing (1) over E^* yields (2). On the other hand, using (b) of Definition 1, we have

$$\sum_{e \in E_i^*(e_i)} m_{\{i'\}}(e_{i'}) = \sum_{e'_i = 1}^{p} m_{\{i'\}}(e_{i'}), \qquad i' \neq i.$$

Again these quantities are zero (Lemma 6.1) and summing (1) over $E_i^*(e_i)$ yields

$$pm_{\{i\}}(e_i) = \sum_{e \in E_i^*(e_i)} (m(e) - m_\varnothing).$$

Finally using (a) of Definition 1 again we obtain

$$\sum_{e \in E_i^*(e_i)} m(e) - pm_\varnothing. \qquad \blacksquare$$

Remark This lemma shows that if the function m on E has Property (1), then it is defined by its values on E^*. Moreover, the components $m_\varnothing$ and

$m_{\{i\}}$ are defined by the same formulas on E^* as on E and the decomposition (1) which is orthogonal when m is considered as a function on E, remains orthogonal when m is considered as a function on E^*. These strong properties are based on the specific structure of an Euler square.

Proof of Theorem 1 Assume that for every $e \in E^*$ we have n independent observations

$$x_e^l (e \in E^*; l = 1, \ldots, n).$$

Then the statistics,

$$\forall e \in E^*, \qquad x(e) = \frac{1}{n} \sum_{l=1}^{n} x_e^l, \qquad s_0 = \sum_{e \in E^*} \left(\sum_{l=1}^{n} (x_e^l - x_e)^2 \right)$$

are independent and sufficient. Denote by x the function defined on E^* by $e \to x(e)$. To every function m on E^* associate the constant $m_\varnothing$ and the functions $m_{\{i\}}$, defined by (2) and (3). Note that here m is a given function on E^* and use only the formulas of Lemma (1). Introduce the quadratic forms

$$i = 1, \ldots, k, \qquad q_i(m) = \|m_{\{i\}}\|^2 = \sum_{e \in E^*} m_{\{i\}}^2(e_i) = p \sum_{e_i = 1}^{p} m_{\{i\}}^2(e_i),$$

$$q_{k+1}(m) = \|m_\varnothing\|^2 = p^2 m_\varnothing^2,$$

$$q_{k+2}(m) = \sum_{e \in E^*} \left(m(e) - m_\varnothing - \sum_{i=1}^{k} m_{\{i\}}(e_i) \right)^2.$$

Using Lemma 1, we have

$$\sum_{i=1}^{k+2} q_i(m) = \sum_{e \in E^*} m^2(e). \tag{4}$$

Assumption (iii) of Theorem 1 shows that $\mathcal{K}$ is now the family of all subsets of K which have at most one element, and we take $\mathcal{K}_0 = \mathcal{K}$. Lemma 1 shows that $(E^*, \mathcal{K}, \mathcal{K}_0)$ is consistent. Let $W_i (i = 1, \ldots, k)$ be the subspace of $\mathcal{X}_{\mathcal{K}}^*$ defined by

$$m(e) = m_{\{i\}}(e_i).$$

(This means that if m belongs to W_i it depends only on the factor i). Let W_{k+1} be the subspace of $\mathcal{X}_{\mathcal{K}}^*$ defined by $m = m_\varnothing$. (This means that if m belongs to W_{k+1}, it does not depend on any factor; i.e., m is constant.) Finally let $W_{k+2} = \mathcal{X}_{\mathcal{K}}^* \cap (W_1 + \cdots + W_k + W_{k+1})^\perp$. Then for every $i = 1, \ldots, k + 2$, $q_i(m)$ is the square of the norm of the projection of m on W_i and the subspaces W_i $(i = 1, \ldots, k + 2)$ are orthogonal and generate $\mathcal{X}_{\mathcal{K}}^*$. Furthermore, since the observations are independent, the inner product on $\mathcal{X}^*$ is

Euclidean for m and for x. The statistics $s_i = q_i(x)$, $i = 1, \ldots, k + 2$, are independent and have distributions,

$$\Gamma\left(\frac{p-1}{2}, q_i(m), \frac{2\sigma^2}{n}\right), \qquad i = 1, \ldots, k,$$

$$\Gamma\left(\frac{1}{2}, q_{k+1}(m), \frac{2\sigma^2}{n}\right), \Gamma\left(\frac{p^2 - pk + k - 1}{2}, \frac{2\sigma^2}{n}\right). \qquad ∎$$

Remark The statistics s_i are computed easily by means of the formulas of Lemma 1. Let $\bar{x}$ be the mean of the $x(e)$, $e \in E^*$, and let x_{ij} be the mean of the $x(e)$ for the elements e in which the factor i has the level j ($i = 1, \ldots, k$, $j = 1, \ldots, p$). Then we have

$$s_i = p \sum_{j=1}^{p} (x_{ij} - \bar{x})^2, \qquad i = 1, \ldots, k,$$

$$s_{k+2} = \sum_{e \in E^*} \left(x(e) - \bar{x} - \sum_{i=1}^{k} (x_{ij} - \bar{x})\right)^2.$$

Finally, we can construct tests of the linear hypotheses W_i, $i = 1, \ldots, k$.

(a) If $n = 1$ and we assume that the function m satisfies (1), then if for some $i = 1, \ldots, k$,

$$\frac{s_i}{s_{k+2}} \geqslant \Phi\left(\alpha; \frac{p-1}{2}, \frac{p^2 - pk + k - 1}{2}\right),$$

we conclude that factor i has an effect on m.

(b) If $n > 1$ and we assume that the function m satisfies (1), then if for some $i = 1, \ldots, k$,

$$\frac{s_i}{s_0 + s_{k+2}} \geqslant \Phi\left(\alpha; \frac{p-1}{2}, \frac{np^2 - pk + k - 1}{2}\right)$$

we conclude that factor i has an effect on m.

9. Introduction to Analysis of Variance (Model II)

Using concepts introduced in Section 6, we now give the basic results for Model II analysis of variance (for a more complete treatment see e.g., Herbach [34] and Lehmann [44, pp. 286–292]). Note that, in this model, the observations are not independent, but there exist some quadratic forms of the observations which are sufficient and independent.

Definition 1 Let E be the set of all levels of the factors, (Section 6). One has a classical *Model II analysis of variance* if for each e belonging to E one has an observation $X(e)$ and if the joint distribution of the family of random variables $X(e)$ is defined by

$$\forall e \in E, \qquad X(e) = \mu + \sum_{\varnothing \neq I \subset K} \sigma_I U^I(e_I),$$

where for every nonempty subset I of K, e_I designates the set of the components of e whose index belongs to I. The constants μ, σ_I $(I \subset K)$ are unknown; the nonobservable random variables $U^I(e_I)$, where I runs through all nonempty subsets of K and e runs through E, are independent standard normal random variables.

In this model, the effect of all factors whose index belongs to I is described by $\sigma_I U^I(e_I)$. Then the standard deviation σ_I depends only on these factors and not on the levels of the factors.

In the special case: $k = 2$, we have

$$X(e_1, e_2) = \mu + \sigma_{\{1\}} U^{\{1\}}(e_1) + \sigma_{\{2\}} U^{\{2\}}(e_2) + \sigma_{\{12\}} U^{\{12\}}(e_1, e_2).$$

Then, denoting

$$v_1 = r, \quad v_2 = s, \quad e_1 = i, \quad e_2 = j, \quad \sigma_{\{1\}} = \sigma_A, \quad U^{\{1\}}(e_1) = A_i,$$

$$\sigma_{\{2\}} = \sigma_B, \quad U^{\{2\}}(e_2) = B_j, \quad \sigma_{\{12\}} = \sigma, \quad U^{\{12\}}(e_1, e_2) = U_{ij},$$

we obtain the usual definition

$$X_{ij} = \mu + \sigma_A A_i + \sigma_B B_j + \sigma U_{ij}; \qquad i = 1, \ldots, r; \;\; j = 1, \ldots, s,$$

where μ, σ_A, σ_B, σ are unknown constants and the $rs + r + s$ random variables A_i, B_j, U_{ij} are independent standard normal random variables.

Model II analysis of variance is based on a fundamental theorem (Theorem 1 below), which we now introduce for the special case: $k = 2$. Using Lemmas 3.1 and 3.2, we can show that the random variable

$$\bar{A} = \frac{1}{r} \sum_i A_i$$

and the random vector

$${}^{t}\mathbf{A} = (A_1 - \bar{A}, \ldots, A_r - \bar{A})$$

are independent. The same result holds for the B_j. Moreover the random variable

$$U_{\varnothing} = \frac{1}{rs} \sum_{ij} U_{ij},$$

the random vectors

$$U_{\{1\}} = \begin{pmatrix} U_{\{1\}}(1) \\ \vdots \\ U_{\{1\}}(r) \end{pmatrix}, \qquad U_{\{2\}} = \begin{pmatrix} U_{\{2\}}(1) \\ \vdots \\ U_{\{2\}}(s) \end{pmatrix},$$

where

$$U_{\{1\}}(i) = \frac{1}{s} \sum_j (U_{ij} - U_\varnothing), \qquad i = 1, \ldots, r$$

$$U_{\{2\}}(j) = \frac{1}{r} \sum_i (U_{ij} - U_\varnothing), \qquad j = 1, \ldots, s,$$

and the random matrix $U_{\{12\}}$ whose general term is given by

$$U_{\{12\}}(i, j) = U_{ij} - U_{\{1\}}(i) - U_{\{2\}}(j) + U_\varnothing,$$

are independent since they are the projections of the sequence $\{U_{ij};\ i = 1, \ldots, r, j = 1, \ldots, s\}$ on the orthogonal subspaces,

$$\mathscr{X}_\varnothing, \mathscr{X}_{\{1\}}, \mathscr{X}_{\{2\}}, \mathscr{X}_{\{12\}}.$$

If we define the real valued statistic $X_\varnothing$, the vector-valued statistics $X_{\{1\}}, X_{\{2\}}$ and the matrix-valued statistic $X_{\{12\}}$ by the same formulas applied to X_{ij}, we obtain

$$X_\varnothing = \mu + \sigma_A \bar{A} + \sigma_B \bar{B} + \sigma U_\varnothing,$$

$$X_{\{1\}} = \sigma_A A + \sigma U_{\{1\}}, \tag{1}$$

$$X_{\{2\}} = \sigma_B B + \sigma U_{\{2\}}, \tag{2}$$

$$X_{\{12\}} = \sigma U_{\{12\}}. \tag{3}$$

Then the statistics $X_\varnothing$, $X_{\{1\}}$, $X_{\{2\}}$, $X_{\{12\}}$ are independent and have the following distributions respectively:

$$N\left(\mu, \frac{\sigma_A^2}{r} + \frac{\sigma_B^2}{s} + \frac{\sigma^2}{rs}\right),$$

$$N\left(O, \left(\sigma_A^2 + \frac{\sigma^2}{s}\right)\mathbf{1}_{\mathscr{X}_{\{1\}}}\right), \quad N\left(O, \left(\sigma_B^2 + \frac{\sigma^2}{r}\right)\mathbf{1}_{\mathscr{X}_{\{2\}}}\right), \quad N(O, \sigma^2 \mathbf{1}_{\mathscr{X}_{\{1,2\}}}), \tag{4}$$

where $\mathbf{1}_{\mathscr{X}_{\{1\}}}, \mathbf{1}_{\mathscr{X}_{\{2\}}}, \mathbf{1}_{\mathscr{X}_{\{1\ 2\}}}$ are the unit quadratic forms on $\mathscr{X}_{\{1\}}, \mathscr{X}_{\{2\}}, \mathscr{X}_{\{1\ 2\}}$, respectively. Then the four statistics,

$$X_\varnothing, S_A = \|X_{\{1\}}\|_{\mathscr{X}}^2 = s \sum_i (X_{\{1\}}(i))^2 = s\|X_{\{1\}}\|_{\mathscr{X}_{\{1\}}}^2,$$

$$S_B = \|X_{\{2\}}\|_{\mathscr{X}}^2 = r \sum_j (X_{\{2\}}(j))^2 = r\|X_{\{2\}}\|_{\mathscr{X}_{\{2\}}}^2,$$

$$S = \|X_{\{12\}}\|_{\mathscr{X}}^2 = \sum_{ij} (X_{\{12\}}(i, j))^2,$$

are independent and have the distributions (4) and

$$\Gamma(\tfrac{1}{2}(r-1), 2\sigma^2 + 2s\sigma_A^2), \Gamma(\tfrac{1}{2}(s-1), 2\sigma^2 + 2r\sigma_B^2), \Gamma(\tfrac{1}{2}(rs-r-s+1), 2\sigma^2),$$

respectively. Moreover, the set of these four statistics is a sufficient statistic.

We now give some tests concerning the unknown parameters. The principle is similar to that used in Model I analysis of variance. The test based on S_A/S is used to test the hypothesis $\sigma_A = 0$ if no additional assumptions are made. If we *assume* that σ_B is zero, the test based on $S_A/(S + S_B)$ is used to test the same hypothesis.

Now, by an analogous argument using notation of Section 6, we can prove the following fundamental theorem.

Theorem 1 *The statistical space corresponding to Definition* 1, *admits as a sufficient statistic the set consisting of the statistics*

$$X_\varnothing, S_I = \|X_I\|_{\mathscr{X}}^2, \qquad (I \neq \varnothing, I \subset K).$$

These statistics are independent. $X_\varnothing$ *has the* $N(\mu, \sum_I(\sigma_I^2/\prod_{j\in I} v_j))$ *distribution and for every nonempty subset* I, S_I *has the distribution*

$$\Gamma\left(\frac{\dim \mathscr{X}_I}{2}, 2\sum_{J\supset I} \sigma_J^2\left(\prod_{j\in \complement J} v_j\right)\right).$$

Proof Consider X and U^I (for every nonempty I) as elements of $\mathscr{X}$. Let U_J^I be the projection of U^I on $\mathscr{X}_J$. The function U^I depends only on the variables e_i, $i \in I$. Lemma 6.3 shows that this function is in $\mathscr{F}_I$ and U_J^I is zero if J is not in I. Thus we have

$$X_\varnothing = \mu + \sum_I \sigma_I U_\varnothing^I,$$

$$X_J = \sum_{J\subset I} \sigma_I U_J^I, \qquad J \neq \varnothing.$$

On the other hand, the random elements U_J^I are independent, since Definition 1 shows that the U^I are independent and Lemma 6.2 shows that the decomposition of X_J is orthogonal. Finally $X_\varnothing$ and X_J ($J \neq \varnothing$) are independent and normal statistics. Moreover, the function U^I has the identity covariance matrix in $\mathscr{F}_I$. Then U_J^I has the covariance matrix

$$\frac{1}{\prod_{j\in I-J} v_j} \mathbf{1}_{\mathscr{X}_J} \tag{5}$$

and X_J has the covariance

$$\left(\sum_{J\subset I} \frac{\sigma_I^2}{\prod_{j\in I-J} v_j}\right)\mathbf{1}_{\mathscr{X}_J} \tag{6}$$

where $\mathbf{1}_{\mathscr{X}_J}$ is the unit quadratic form in $\mathscr{X}_J$.

The set of the statistics

$$X_\varnothing, \quad S_I = \|X_I\|^2_{\mathscr{X}} = \Big(\prod_{j\in \complement I} v_j\Big)\|X_I\|^2_{\mathscr{X}_I} \qquad (I \neq \varnothing, I \subset K) \tag{7}$$

is a sufficient statistic and these statistics are independent. The proof is completed by using the elementary identity

$$\Big(\prod_{j\in \complement I} v_j\Big)\Big(\sum_{J\supset I} \frac{\sigma_J^2}{\prod_{j\in J-I} v_j}\Big) = \sum_{J\supset I} \sigma_J^2\Big(\sum_{j\in \complement J} v_j\Big).$$

■

10. Generalized Linear-Normal Statistical Spaces

In this section we discuss statistical spaces which are fundamental for multivariate analysis. They are direct generalizations of statistical spaces used in Section 2.

Let $\mathscr{X}$ be a finite-dimensional linear space, let Λ be a quadratic form on $\mathscr{X}$ and let Σ be a matrix in $\mathscr{S}_p^+$, where p is a given integer. We designate by $\Lambda \otimes \Sigma$ the quadratic form on $\mathscr{X}^p$, defined by

$$(x_1, \ldots, x_p) \to \mathrm{Tr}(\Sigma M_\Lambda), \tag{1}$$

where M_Λ is the symmetric matrix of order p whose general term is $f_\Lambda(x_i, x_j)$ and f_Λ is the bilinear form associated with Λ. Note that this definition is consistent with the definitions in Chapter 8 (see Exercise 30).

Definition 1 We call

$$[\Omega^p, \mathscr{B}_{\Omega^p}; N(m, \Lambda \otimes \Sigma), m \in V^p, \Sigma \in \mathscr{S}_p^+]$$

a *linear-normal* statistical space of *order p*, where Ω is a finite-dimensional linear space, $\mathscr{B}_{\Omega^p}$ is the Borel subfield of Ω^p, V is a linear subspace of Ω and $V \neq \Omega$, Λ is a given positive-definite quadratic form on Ω^* the dual space of Ω, and the inner product on Ω is defined by Λ^{-1}.

We have seen (Section 2) that the normal distribution $N(m, \Lambda \otimes \Sigma)$ is defined by m and the quadratic form $\Lambda \otimes \Sigma$. If the random element $(w_1, \ldots, w_p)$ has the $N(O, \Lambda \otimes \Sigma)$ distribution, then for every x and y in Ω^* the following argument shows that we have

$$E(x(w_i)\,y(w_j)) = \sigma_{ij} f_\Lambda(x, y), \qquad i, j = 1, \ldots, p,$$

where $\Sigma = \{\sigma_{ij}\}$. The bilinear form associated with $\Lambda \otimes \Sigma$ is defined by

$$(x_1, \ldots, x_p; y_1, \ldots, y_p) \to \sum_{i,j=1,\ldots,p} \sigma_{ij} f_\Lambda(x_i, y_j), \qquad (x_i, y_j \in \Omega^*).$$

Thus, the previous expectation is equal to this bilinear form if $x_i = x$, $y_i = y$ and the other components are zero. Definition 1 introduced an inner product on Ω. We introduce on Ω^p the inner product

$$\langle w_1, \ldots, w_p | w'_1, \ldots, w'_p \rangle = \langle w_1 | w'_1 \rangle + \cdots + \langle w_p | w'_p \rangle.$$

If M designates the matrix which defines the quadratic form Λ^{-1} with respect to a given basis in Ω, then the previous inner product on Ω^p is defined by the matrix $M \otimes \mathbb{1}_p$.

Now let V be a linear subspace of Ω. It is easy to see that

$$\forall z = (z^1, \ldots, z^p) \in \Omega^p, \qquad z_{V^p} = (z^1_V, \ldots, z^p_V).$$

Then, if we denote by P the projection matrix in V with respect to a basis of Ω, we have

$$z_{V^p} = Pz,$$

where we have written z as a matrix. This relation will be useful since in frequently used cases P is given by the formulas of Theorem 2.1. Finally, for every $z = (z_1, \ldots, z_p) \in \Omega^p$, we designate by $Q(z)$ the matrix in $\mathscr{S}_p$ whose general term is $\langle z_i | z_j \rangle$ and we have

$$Q(z) = Q(z_{V^p}) + Q(z_{(V^\perp)^p}). \tag{2}$$

Theorem 1 *Let $\mathscr{X}$ be a finite-dimensional linear space and let X be a random element having values on $\mathscr{X}^p$ with the nonsingular $N(m, \Lambda \otimes \Sigma)$ distribution. If the inner product on $\mathscr{X}$ is defined by Λ^{-1} and if $F_1, \ldots, F_k$ are k orthogonal subspaces of $\mathscr{X}$, then*

(i) *the random elements $X_{F_j^p}$ $(j = 1, \ldots, k)$ are independent and have $N(m_{F_j^p}, \Lambda_{F_j} \otimes \Sigma)$ distributions, where the matrix Λ_{F_j} was used in Section 3 for the case, $p = 1$ (see Theorem 3.1).*

(ii) *the random matrices $Q(X_{F_j^p})$ $(j = 1, \ldots, k)$ are independent and have $\Gamma_p(\frac{1}{2} \dim F_j, Q(m_{F_j^p}), 2\Sigma)$ distributions.*

Proof First, we establish the following:

Lemma *Let $X[p, q]$ be a random matrix having the $N(m, \Lambda \otimes \Lambda')$ distribution and let M be a matrix in $\mathscr{S}_p$. If Λ is nonsingular and if all the r nonzero eigenvalues of ΛM are equal to s, then the matrix ${}^tXMX$ is distributed as $\Gamma_q(\frac{1}{2}r, {}^tmMm, 2s\Lambda')$.*

Proof of the Lemma Lemma 7.2.1 shows that there exists a nonsingular square matrix A of order p such that

$$\Lambda = A\,{}^tA, \qquad {}^tAMA = s\begin{pmatrix} \mathbb{1}_r & 0 \\ 0 & 0 \end{pmatrix}.$$

Then we can write

$$X = m + AU,$$

where the random matrix U has the $N(O, \mathbb{1}_p \otimes \Lambda')$ distribution. Now let m' and u' be the matrices obtained from $A^{-1}m$ and U respectively, by keeping only the r first rows. Then we have

$$\begin{aligned} {}^{t}XMX &= {}^{t}(m + Au)M(m + Au) = s\,{}^{t}(A^{-1}m + u)\begin{pmatrix} \mathbb{1}_r & 0 \\ 0 & 0 \end{pmatrix}(A^{-1}m + u) \\ &= s\,{}^{t}(m' + u')(m' + u'). \end{aligned}$$

But $m' + u'$ has the $N(m', \mathbb{1}_r \otimes \Lambda')$ distribution. The proof is completed by using Theorem 8.10.1. ∎

To prove Theorem 1, choose a basis in $\mathscr{X}$ and consider the case where $\mathscr{X} = \mathbb{R}^n$. The proof consists in applying Theorem 3.1 to each column of X. Let B be a matrix, where ${}^{t}BB = \Sigma$. We can write

$$X = m + YB,$$

where Y has the $N(O, \Lambda \otimes \mathbb{1}_p)$ distribution. This means that Y^i $(i = 1, \ldots, p)$, the columns of Y, are independent random vectors having $N(O, \Lambda)$ distributions. Apply Theorem 3.1 to each Y^i. Then, letting P_j be the projection matrix on F_j, we see that the random vectors,

$$Y^i_{F_j} = P_j Y^i, \qquad i = 1, \ldots, p, \quad j = 1, \ldots, k,$$

are independent and have $N(O, \Lambda_j)$ distributions. Therefore, the matrices

$$Y_j = (Y^1_{F_j}, \ldots, Y^p_{F_j}) = P_j Y, \qquad j = 1, \ldots, k$$

are independent and have $N(O, \Lambda_{F_j} \otimes \mathbb{1}_p)$ distributions. This completes the proof of (i) since

$$X_{F^p_j} = P_j X = P_j m + Y_j B.$$

Now we have

$$Q(z) = {}^{t}z\Lambda^{-1}z.$$

Then

$$Q(X_{F^p_j}) = {}^{t}X\,{}^{t}P_j\Lambda^{-1}P_jX.$$

The distributions of these matrices are obtained by using the lemma with

$$\Lambda' = \Sigma, \qquad M = {}^{t}P_j\Lambda^{-1}P_j.$$

But the matrix $\Lambda\,{}^{t}P_j\Lambda^{-1}P_j$ is equal to P_j and this matrix has $\dim F_j$ nonzero eigenvalues which are all equal to 1. ∎

Theorem 2 *The statistic*

$$(x_{VP}, Q(x_{(V^{\perp})P})) \tag{3}$$

is sufficient and complete. Moreover, x_{VP} *and* $Q(x_{(V^{\perp})P})$ *are independent and have the following distributions respectively,*

$$N(m, \Lambda_V \otimes \Sigma), \qquad \Gamma_p(\tfrac{1}{2} \dim V^{\perp}, 2\Sigma). \tag{4}$$

Proof The logarithm of the likelihood function differs only by a constant from the function,

$$x \in \Omega^p \to \mathrm{Tr}(\Sigma^{-1} M_{\Lambda^{-1}}(x - m)), \tag{5}$$

where

$$M_{\Lambda^{-1}}(x - m) = Q(x - m).$$

Now (5) implies that

$$Q(x - m) = Q(x_{VP} - m) + Q(x_{(V^{\perp})P})$$

and the factorization theorem shows that (3) is a sufficient statistic. Applying Theorem 1 with $F_1 = V$ and $F_2 = V^{\perp}$, we obtain (4).

Finally, to prove that (3) is complete, choose an orthonormal basis of Ω such that the first q ($q = \dim V$) vectors of this basis are a basis of V. Use the corresponding basis on Ω^p and identify Λ^{-1}, x, x_{VP}, $\cdots$ with the corresponding matrices. Then,

$$\begin{aligned} Q(x_{VP} - m) &= {}^t(x_{VP} - m)\Lambda^{-1}(x_{VP} - m) \\ &= {}^t x_{VP}\Lambda^{-1}x_{VP} - {}^t x_{VP}\Lambda^{-1}m - {}^t m\Lambda^{-1}x_{VP} + {}^t m\Lambda^{-1}m. \end{aligned}$$

Now (5) is equal to

$$\mathrm{Tr}(\Sigma^{-1}\mathrm{Q(x)}) - 2\,\mathrm{Tr}(\Sigma^{-1}\,{}^t m\Lambda^{-1}x_{VP}) + \mathrm{Tr}(\Sigma^{-1}Q(m)).$$

Let Θ and x^*_{VP} be the $[q, p]$ matrices defined by the first q rows of the components of $\Sigma^{-1}\,{}^t m$ and of x_{VP} respectively. Then (4) is equal to

$$\mathrm{Tr}(\Sigma^{-1}Q(x)) - 2\,\mathrm{Tr}(\Theta\Lambda^{-1}x^*_{VP}) + \mathrm{Tr}(\Sigma^{-1}Q(m))$$

and the statistical space induced by the statistic $(Q(x_{(V^{\perp})P}), x^*_{VP})$ is a standard exponential statistical space whose parameters are defined by the elements of Θ and by the elements of $\Sigma^{-}1$ which are not below the diagonal. ∎

We conclude this section by noting that if W is a linear subspace of V and T the orthogonal complement of W in V, the test of the hypothesis $\{m_{TP} = 0\}$ is based on the statistic,

$$Z = \frac{\det[Q(x_{(V^{\perp})P})]}{\det[Q(x_{TP}) + Q(x_{(V^{\perp})P})]}.$$

Under the null hypothesis, the distribution of this statistic is given in Theorem 8.9.6.

Exercises

1. Let q and q' be two quadratic forms on $\mathbb{R}^n$. We say that $q \geqslant q'$ if for every $x \in \mathbb{R}^n$, we have

$$q(x) \geqslant q'(x).$$

Show that this condition is equivalent to saying that $q - q'$ is a nonnegative definite quadratic form. Now let x and y be two matrices of the same order. Show that if $x\,{}^t x$ is nonsingular,

$$y\,{}^t x(x\,{}^t x)^{-1} x\,{}^t y \leqslant y\,{}^t y.$$

2. Using Theorem 2.1 and Lagrange multipliers, find for a given x the minimum value of ${}^t(x - A\theta)M(x - A\theta)$ with respect to θ, subject to $B\theta = 0$.

3. Using Theorem 2.1, compute x_V in the two special cases

(a) $B = 0$ and S is nonsingular;

(b) $A = \mathbb{1}_n$ and $BM^{-1}\,{}^t B$ is nonsingular.

4. Let Q be a positive-definite matrix of order n and let $B[p, n]$ $(p \leqslant n)$ be a matrix whose rows are linearly independent. Show that the matrix $BQ\,{}^t B$ is positive definite.

5. Show that the set of samples from a $N(m, \sigma^2)$ distribution forms a linear-normal statistical space.

6. Let $\mathscr{X}$ be a finite-dimensional linear space, and W a linear subspace of $\mathscr{X}$. Let X be a random vector taking values in $\mathscr{X}$, having the nonsingular $N(m, \Lambda)$ distribution. Introduce on $\mathscr{X}$ the inner product defined by Λ^{-1} and consider the linear space $\mathscr{X}/W^{\perp}$, which is the space of the equivalence class of vectors of $\mathscr{X}$ having the same projection on W. Define a "natural" inner product on $\mathscr{X}/W^{\perp}$. Show that there exists a random vector $\tilde{X}_W$ on $\mathscr{X}/W^{\perp}$ which corresponds to X_W, compute its distribution, show that its covariance is nonsingular and find the relationship between this covariance and the inner product on $\mathscr{X}/W^{\perp}$.

7. Prove Theorem 3.1 by choosing an orthogonal basis in each subspace, $F_1, \ldots, F_k$. Give another direct argument to prove this theorem by using Exercise 6.

8. Compute the correlation coefficients between the random variables

$$\frac{\|X_{F_i}\|^2}{\|X_{F_1}\|^2} \quad \text{and} \quad \frac{\|X_{F_j}\|^2}{\|X_{F_1}\|^2} \qquad i, j = 2, \ldots, n, \quad i \neq j$$

defined in Theorem 3.1, when m_{F_1} is zero. Find the minimum of these coefficients with respect to m.

9. Prove the Corollary of Theorem 3.1.

10. Using the Corollary of Theorem 3.1, prove Fisher's theorem concerning the sample mean and the sample covariance of a normal sample.

11. Let F' be a subspace of V, where V is given in Definition 2.1. Find the distribution of the statistic

$$X \to \|X_{F'}\|^2/\|X_F\|^2,$$

where F is a subspace of Ω orthogonal to V. Show that the result remains valid if F is the subspace spanned by k random vectors independent of each other and of X.

12. Let x_i^j ($i = 1, \ldots, r$; $j = 1, \ldots, n_i$; $n_1 + \cdots + n_r > r$) be independent observations having $N(m_i, \sigma^2)$ distributions. The parameters $m_1, \ldots, m_r, \sigma$ are unknown. Write the statistical space corresponding to this configuration and find the test of the linear hypothesis $\{m_1 = m_2 = \cdots = m_r\}$.

13. Let x_i ($i = 1, \ldots, n$) be n independent observations having $N(\alpha + \beta t_i, \sigma^2)$ distributions, where the t_i are given and the parameters α, β, σ are unknown. Write the statistical space corresponding to this configuration and find the test of the linear hypothesis $\{\beta = 0\}$.

14. Assume that the subspaces $\{W_j, j = 1, \ldots, k\}$, of Definition 4.1, are orthogonal. Find the joint distribution of the statistics $\|x_{V^\perp}\|^2$, $\|x_{W_1}\|^2, \ldots, \|x_{W_k}\|^2$ and use it to describe a way of computing the power function of the strategy given in Definition 4.1. How can Cochran's theorem be used to compute the above statistics?

15. Using the notation of Theorem 5.1, let γ be a real number which is greater than $-\dim V^\perp$. Find an unbiased estimator and a confidence interval of minimum length for σ^γ.

16. Refer to Theorem 5.1. Let f be a polynomial in σ and the components of m. Find the estimator of $f(m, \sigma)$ which has minimum variance.

17. Give the simplified results of Theorem 5.3 in the special case where $B = 0$.

18. Consider the statistical space $[\mathbb{R}^n, \mathscr{B}_{\mathbb{R}^n}, \mathscr{P}]$, where $\mathscr{P}$ is a family of distributions on $\mathbb{R}^n$, such that

(1) the mean of each distribution is in a given subspace V of $\mathbb{R}^n$;
(2) the covariance of each distribution is equal to $\sigma^2\Lambda$, where Λ is a given covariance and σ is an unknown parameter.

Show that for every inner product given on $\mathbb{R}^n$, the statistic $x \to x_V$ is an unbiased estimate of m. Using Exercise 1, show that if we choose on $\mathbb{R}^n$

the inner product defined by Λ^{-1}, the statistic $x \to x_V$ is an unbiased estimator of m whose variance is less than the variance of any other linear unbiased estimator of m.

19. If in a linear normal statistical space the variance σ is in fact known, how can the problem of testing linear hypotheses or the estimation of linear functions be simplified?

20. Apply Definition 6.1 to the special cases, $k = 1, 2, 3$. In each case give the formulas defining the functions m_I.

21. Let $U_1, \ldots, U_k$ be k independent random variables and f a measurable mapping from $\mathbb{R}^k$ to $\mathbb{R}$ such that the random variable $X = f(U_1, \ldots, U_k)$ is integrable. Show that for every subset I of $K = \{1, \ldots, k\}$ there exists a mapping f_I from $\mathbb{R}^k$ to $\mathbb{R}$ such that

(1) $f_I(U_1, \ldots, U_k)$ depends only on the variables U_j for $j \in I$

(2) $f = \sum_{I \subset K} f_I$

(3) $\forall j \in I, \quad E(f_I(\{U_j, j \in I\}) | \{U_i, i \in I - j\}) = 0.$

22. Assume in the methodology of Section 7 that for every $e \in E^*$ one has $n(e)$ observations on X, designated by $x_i(e)$ $(i = 1, \ldots, n(e))$ and that the correlation coefficient, $\rho_{ii'}(e, e')$ between any two of these observations is given. Show that this problem can be considered as an analysis of variance problem. Write the corresponding statistical space; find a sufficient statistic; and, finally, obtain a statistical space analogous to that for Problems 2 and 3 of Section 7.

23. Using Cochran's theorem verify directly that the spaces W_1, W_2, W_3, W_4, in a factorial experimental design with two factors (Section 8), are orthogonal. Compute the power function of the linear tests which are used in this design.

24. Consider the problem of analysis of variance on an Euler square (Section 8) and prove directly that

$$\sum_{i=1}^{k+2} q_i(m) = \sum_{e \in E^*} m^2(e).$$

25. Compute the correlation coefficient between $X(e)$ and $X(e')$ $(e, e' \in E, e \neq e')$ for Model II concepts given in Definition 9.1. Deduce from this the covariance of the set of variables $\{X(e), e \in E\}$.

26. Show that the set of all samples from the multidimensional normal distribution forms a generalized linear-normal statistical space.

27. Let x_i, $i = 1, \ldots, n$ be n independent random vectors of $\mathbb{R}^p$ having $N(Ba_i, \Sigma)$ distributions. If a_i are given vectors of $\mathbb{R}^p$, and if the matrices B

and Σ are unknown ($\Sigma \in \mathcal{S}_p^+$), show that the corresponding statistical space is a linear normal statistical space of order p.

28. Give the results corresponding to those in Theorem 10.1 for the case in which $\mathcal{X} = \mathbb{R}^n$ and the subspaces F_j are defined by their projection matrices.

29. Find the conditions on m necessary for the results of Lemma 10.1 to remain valid when Λ is singular.

30. Using notation introduced in the beginning of Section 10, show that if $\mathcal{X} = \mathbb{R}^n$, the matrix $\Lambda \otimes \Sigma$ defines the quadratic form $\Lambda \otimes \Sigma$. Then show that, in the general case, if we choose a basis in $\mathcal{X}$, the matrix which defines the quadratic form $\Lambda \otimes \Sigma$ is the direct product of the matrix corresponding to Λ and the matrix Σ. Deduce from this that the definition of the quadratic form $\Lambda \otimes \Sigma$ does not depend on the choice of the basis in $\mathcal{X}$.

CHAPTER **10**

Exponential Statistical Spaces

Exponential statistical spaces are very useful in mathematical statistics and their properties are based on some results concerning the Laplace transform. First we recall some properties which will be needed.

1. Laplace Transform of a Measure

Let $\mathbb{C}^s$ be the s-fold cartesian product of the complex plane $\mathbb{C}$. The open (or closed) *polydisk* [13, p. 203] having center $a = (a_1, \ldots, a_s)$ and radius $r = (r, \ldots, r_s)$ is defined by

$$|z_j - a_j| < r_j \qquad \{\text{or} \quad |z_j - a_j| \leqslant r_j\}, \qquad j = 1, \ldots, s.$$

In this chapter we shall use the following notation, where the set of the non-negative integers is designated by $\mathbb{N}$:

$$\nu = (n_1, \ldots, n_s) \in \mathbb{N}^s, \qquad \nu! = n_1! \cdots n_s!, \qquad |\nu| = n_1 + \cdots + n_s,$$

$$z = (z_1, \ldots, z_s) \in \mathbb{C}^s, \qquad z^\nu = z_1^{n_1} z_2^{n_2} \cdots z_s^{n_s}.$$

Finally, $\mathrm{Re}(z)$ designates the real part of z.

Definition 1 [13, p. 203] Let f be a mapping from $\mathbb{C}^s$ into C; f is *analytic* on an open set D of $\mathbb{C}^s$ if for every point $a \in D$ there exists an open polydisk P, contained in D, which has a as center, such that,

$$\forall z \in P, \qquad f(z) = \sum_{\nu \in \mathbb{N}^s} c_\nu (z - a)^\nu,$$

and the series is absolutely convergent.

Hartogs' theorem (Hormander [28, p. 28] or Dieudonné [13, p. 226]) shows that such a function is analytic if and only if it is analytic with respect to each variable.

Theorem 1 [13, p. 208] *Let A be an open and connected set of $\mathbb{C}^s$. Let f and g be two analytic functions from A into $\mathbb{C}$. Let U be an open subset of A and let b be a point of U. Then if f and g are equal on $U \cap (b + \mathbb{R}^s)$, they are equal on A.*

Specifically, if $A \cap \mathbb{R}^s$ is nonempty, then every analytic function on A is defined by its values on $A \cap \mathbb{R}^s$. In particular, if an analytic function on A vanishes on $A \cap \mathbb{R}^s \neq \varnothing$, then it vanishes on A.

Definition 2 Let m be a positive measure on $(\mathbb{R}^s, \mathscr{B}_{\mathbb{R}^s})$. We denote by D_m the convex set of $\mathbb{R}^s$ defined by

$$\theta \in D_m \quad \Leftrightarrow \quad \int_{\mathbb{R}^s} e^{-\langle \theta | x \rangle}\, dm(x) < \infty.$$

If D_m is nonempty, the function which is defined by

$$L_m(z) = \int_{\mathbb{R}^s} e^{-\langle z | x \rangle}\, dm(x)$$

is called the *Laplace transform* of m on the following set of $\mathbb{C}^s$

$$B_m = \{\mathrm{Re}(z) \in D_m\}.$$

Remark Since the measure m is positive, the convexity of the exponential function implies that D_m is a convex set. On the other hand, we have

$$\forall z \in B_m, \qquad |e^{-\langle z | x \rangle}| = e^{-\langle \mathrm{Re}(z) | x \rangle},$$

and then $L_m(z)$ is defined for any z belonging to B_m and the definition is consistent. Obviously in the special case $s = 1$, D_m is an interval and the boundedness of this interval depends on the support of m. In the general case, if there exists a value of θ such that

$$m(\{\langle \theta | x \rangle \geqslant 0\}) = 0,$$

then, if D_m is not empty, it contains all the half-lines

$$\forall \lambda \in \mathbb{R}^+, \qquad \forall \theta_0 \in D_m, \qquad \{\theta_0 - \lambda\theta\}.$$

Definition 3 Let m be a measure on $(\mathbb{R}^s, \mathscr{B}_{\mathbb{R}^s})$ and let m^+ and m^- be its positive and negative parts respectively. If one has

$$D_m = D_{m^+} \cap D_{m^-} \neq \varnothing,$$

such a measure is called a smooth measure and we designate by $\tilde{\mathscr{M}}_s$ the family of these measures on $(\mathbb{R}^s, \mathscr{B}_{\mathbb{R}^s})$. For every smooth measure m we define the Laplace transform of m by

$$L_m(z) = L_{m^+}(z) - L_{m^-}(z).$$

Remark The measure m is smooth if and only if there exist two probability distributions P_1 and P_2 on $(\mathbb{R}^s, \mathscr{B}_{\mathbb{R}^s})$, two nonnegative constants A and B and a vector θ of $\mathbb{R}^s$, such that

$$dm = Ae^{\langle \theta | x \rangle}\, dP_1 - Be^{\langle \theta | x \rangle}\, dP_2.$$

For example, the Lebesgue measure on the positive cone of $\mathbb{R}^s$ is a smooth measure. However, the Lebesgue measure on the whole space $\mathbb{R}^s$ is not smooth.

Theorem 2 [44, p. 52] *Let m be a smooth measure. If the interior set D_m^0 is nonempty, then $L_m(z)$ is an analytic function on*

$$B_m^0 = \{\mathrm{Re}(z) \in D_m^0\}$$

and, for every $z_0 \in B_m^0$,

$$z \in B_m^0, \qquad L_m(z) = \sum_{\nu \in \mathbb{N}^s} \frac{(-1)^{|\nu|}}{\nu!}(z - z_0)^\nu \cdot \int_{\mathbb{R}^s} e^{-\langle z_0 | x \rangle} x^\nu \, dm(x).$$

Proof Using the above discussion, we can reduce the theorem to the special case, where $s = 1$ and m is a probability distribution. But in this case, the following result is an elementary one. Let X be a random variable, such that $e^{-\theta X}$ is integrable for every θ belonging to an open interval which contains $\{0\}$. Then X has moments of any order and one has

$$\forall z \in \mathbb{C}, \quad \nu \in \mathbb{N}, \qquad \left.\frac{d^\nu E(e^{-zX})}{dz^\nu}\right|_{z=0} = -E(X^\nu). \qquad \blacksquare$$

Theorem 3 *If $L_m(z)$ vanishes for all real z belonging to some open subset of D_m, then m is zero.*

Proof The function $L_m(z)$ is an analytic function on B_m^0. This set is convex and thus is a connected set. Moreover, $L_m(z)$ vanishes on an open subset of D_m. Then, Theorem 1 implies that $L_m(z)$ vanishes on B_m^0. Let θ_0 be an interior point of D_m, then

$$L_{m^+}(\theta_0) = L_{m^-}(\theta_0) = K > 0.$$

The two measures defined by

$$dP^+ = \frac{e^{-\langle\theta_0|x\rangle}}{K} dm^+, \qquad dP^- = \frac{e^{-\langle\theta_0|x\rangle}}{K} dm^-$$

are two probability distributions which have the same characteristic function, i.e., $\varphi_{P^+} = \varphi_{P^-}$, since

$$\forall t \in \mathbb{R}^s, \qquad K\varphi_{P^+}(t) = L_{m^+}(\theta_0 + it) = L_{m^-}(\theta_0 + it) = K\varphi_{P^-}(t).$$

Then, since $e^{-\langle\theta_0|x\rangle}$ is always positive, m is zero. ∎

Remark The previous proof shows that if there exists $\theta_0 \in D_m$, such that $L_m(\theta_0 + it)$ is zero for every t belonging to $\mathbb{R}^s$, then m is zero.

Theorem 4 *Let m_1 and m_2 be two smooth measures such that*

$$D_{m_1} \cap D_{m_2} \neq \varnothing.$$

Then, there exists a unique smooth measure denoted by $m_1 * m_2$, called the convolution *of m_1 and m_2, such that*

$$L_{m_1 * m_2}(z) = L_{m_1}(z)L_{m_2}(z), \qquad \forall z \in B_{m_1} \cap B_{m_2}.$$

Proof Using Definition 3 it is sufficient to prove this theorem in the case of two positive measures. But in the case of two probability distributions we see easily that this definition of convolution is the classical one. ∎

2. Analytical Properties of Exponential Statistical Spaces

Definition 1 We say that the dominated statistical space $[\mathbb{R}^k, \mathscr{B}_{\mathbb{R}^k}; p_\theta, \theta \in \Theta]$ is *exponential* if

(i) the support of p_θ does not depend on θ, i.e.,

$$\forall \theta \in \Theta \qquad E = \{x \in \mathbb{R}^k : p_\theta(x) > 0\},$$

(ii) the linear space of functions on E, generated by the family of functions on E,

$$\log \frac{p_\theta(x)}{p_{\theta_0}(x)}, \qquad \theta \in \Theta,$$

where $\theta_0 \in \Theta$, has finite dimension.

This definition means that $p_\theta(x)$ has the form,

$$C(\theta)\exp\left\{\sum_{j=1}^{s} Q_j(\theta)T_j(x)\right\}h(x).$$

The set of such distributions is usually called the *exponential family.*

Definition 2 Let m be a positive smooth measure on $(\mathbb{R}^s, \mathscr{B}_{\mathbb{R}^s})$. Any statistical space such that

$$\left[\mathbb{R}^s, \mathscr{B}_{\mathbb{R}^s}; \frac{dP_\theta}{dm} = \frac{e^{\langle\theta|x\rangle}}{L_m(-\theta)}, \theta \in \Theta \subset D'_m\right],$$

where D'_m is the set symmetric to D_m with respect to the origin, is called a *standard* exponential statistical space associated with m.

Theorem 1 *Let*

$$\left[\mathbb{R}^s, \mathscr{B}_{\mathbb{R}^s}; \frac{dP_\theta}{dm} = \frac{e^{\langle\theta|x\rangle}}{L_m(-\theta)}, \theta \in \Theta \subset D'_m\right],$$

be a standard exponential statistical space, where m is a positive smooth measure. If the interior set of Θ is nonempty, this statistical space is complete.

Proof Let f be a function on $\mathbb{R}^s$ such that

$$\int fe^{\langle\theta|x\rangle}\,dm = 0, \qquad \forall\theta \in \Theta.$$

Let θ_0 be a point of Θ and let m' be the measure defined by

$$dm' = fe^{\langle\theta_0|x\rangle}\,dm.$$

Then m' is a smooth measure and we have

$$\theta_0 - \text{Re}(z) \in \Theta \quad \Rightarrow \quad L_{m'}(\text{Re}(z)) = 0.$$

Therefore, using Theorem 1.3, m' is zero and thus f is zero almost everywhere. ∎

Theorem 2 *Let T be a random vector on $(\mathbb{R}^s, \mathscr{B}_{\mathbb{R}^s})$ having the distribution P_θ whose density is*

$$\frac{dP_\theta}{dm} = \frac{1}{L_m(-\theta)} \cdot e^{\langle\theta|x\rangle},$$

where m is a positive smooth measure and θ a given point of D'_m. Then the *characteristic function of T is equal to*

$$\forall t \in \mathbb{R}^s, \qquad \varphi_T(t) = \frac{L_m(-it - \theta)}{L_m(-\theta)}.$$

Moreover, if θ is an interior point of D_m, then T has moments of any order. In particular,

$$E(T) = \operatorname{grad}_\theta \log L_m(-\theta),$$

$$\Lambda_T = \left\{\frac{\delta^2}{\delta\theta_i\,\delta\theta_j} \log L_m(-\theta);\ i, j = 1, \ldots, s\right\}.$$

The proof of this theorem is left as an exercise.

Theorem 3 *Let U and V be the vectors defined by the first h components of T and by the k last components of T, respectively ($h + k = s$). Express θ as (λ, μ) in the same way. Then there exist measures m_λ and q_v defined on $(\mathbb{R}^k, \mathscr{B}_{\mathbb{R}^k})$ and $(\mathbb{R}^h, \mathscr{B}_{\mathbb{R}^h})$, respectively, such that*

(i) *the distribution P_V^θ of V is given by*

$$dP_V^\theta = \frac{1}{L_m(-\theta)}\, e^{\langle\mu|v\rangle}\, dm_\lambda;$$

(ii) *the conditional distribution of U given V does not depend on μ and is defined by*

$$dP_{U/V}^{\lambda, v} = \frac{1}{L_{q_v}(-\lambda)}\, e^{\langle\lambda|u\rangle}\, dq_v.$$

Proof Without loss of generality we can restrict ourselves to the case where m is a probability distribution. Let (X, Y) be a random vector having this distribution. We have

$$\begin{aligned}
\forall A \in \mathscr{B}_{\mathbb{R}^k}, \qquad P_V^\theta(A) &= \frac{1}{L_m(-\theta)} E(1_A \exp(\langle \lambda X + \mu Y\rangle)) \\
&= \frac{1}{L_m(-\theta)} E[1_A\, e^{\langle\mu|Y\rangle} E(e^{\langle\lambda|X\rangle} \mid Y)] \\
&= \int_A \frac{e^{\langle\mu|Y\rangle}}{L_m(-\theta)} \cdot E(e^{\langle\lambda|X\rangle} \mid Y)\, dm_Y,
\end{aligned}$$

where expectations are taken with respect to m. Then (i) is proved by choosing m_λ, defined by

$$\frac{dm_\lambda}{dm_Y} = E(e^{\langle\lambda|X\rangle} \mid Y) = L_{m^Y_{X|Y}}(-\lambda).$$

Recall that m_Y and $m^y_{X|Y}$ designate the marginal distribution of Y and the conditional distribution of X given Y, respectively.

To prove (ii), choose $m^v_{X|Y}$ for the distribution q_v and verify the identity of conditional probability. Then it must be shown that

$$\forall A \in \mathscr{B}_{\mathbb{R}^h}, \quad C \in \mathscr{B}_{\mathbb{R}^k}, \qquad P_\theta(A \times C) = \int_C \left[\int_A \frac{e^{\langle \lambda | u \rangle}\, dm^v_{X|Y}}{L_{m^v_{X|Y}}(-\lambda)} \right] dP^\theta_V,$$

or that

$$P_\theta(A \times C) = \int_{A \times C} \frac{e^{\langle \lambda | u \rangle}}{L_{m^v_{X|Y}}(-\lambda)} \cdot \frac{1}{L_m(-\theta)} e^{\langle \mu | v \rangle}\, dm^v_{X|Y}\, dm_\lambda(v).$$

Replacing m_λ by its value, we obtain

$$P_\theta(A \times C) = \int_{A \times C} \frac{e^{\langle \lambda | u \rangle}}{L_{m^v_{X|Y}}(-\lambda)} \cdot \frac{1}{L_m(-\theta)} e^{\langle \mu | v \rangle} L_{m^v_{X|Y}}(-\lambda)\, dm_Y\, dm^v_{X|Y}$$

$$= \int_{A \times C} \frac{e^{\langle \lambda | u \rangle + \langle \mu | v \rangle}}{L_m(-\theta)}\, dm = \int_{A \times C} dP_\theta,$$

and the proof is completed, since by assumption one has

$$dm_Y\, dm^y_{X|Y} = dm. \qquad ∎$$

Theorem 4 *Let T and T' be two independent random vectors on $(\mathbb{R}^s, \mathscr{B}_{\mathbb{R}^s})$, having probability density functions,*

$$\frac{dP_T}{dm} = \frac{e^{\langle \theta | x \rangle}}{L_m(-\theta)}, \qquad \frac{dP_{T'}}{dm'} = \frac{e^{\langle \theta | x \rangle}}{L_{m'}(-\theta)},$$

where m and m' are two positive smooth measures on $(\mathbb{R}^s, \mathscr{B}_{\mathbb{R}^s})$ and $\theta \in D'_m \cap D'_{m'}$. Then the random vector $Z = T + T'$ has the probability density function

$$\frac{dP_Z}{dm * m'} = \frac{e^{\langle \theta | x \rangle}}{L_m(-\theta) L_{m'}(-\theta)}.$$

Proof We have

$$\forall t \in \mathbb{R}^s, \qquad \varphi_Z(t) = \varphi_T(t)\varphi_{T'}(t) = \frac{L_m(-it - \theta)}{L_m(-\theta)} \cdot \frac{L_{m'}(-it - \theta)}{L_{m'}(-\theta)}$$

$$= \frac{L_{m*m'}(-it - \theta)}{L_{m*m'}(-\theta)}.$$

The proof is completed, since the characteristic function φ_Z determines the distribution of Z. ∎

3. Sufficient Statistics on an Exponential Statistical Space

Theorem 1 *On the following exponential statistical space*

$$\left[\mathbb{R}^k, \mathscr{B}_{\mathbb{R}^k}; C(\theta)\exp\left\{\sum_{j=1}^{s} Q_j(\theta)T_j(x)\right\}h(x),\ \theta\in\Theta\right]^n$$

the statistic T taking values in $(\mathbb{R}^s, \mathscr{B}_{\mathbb{R}^s})$, which is defined by

$$T(x_1,\ldots,x_n) = \left\{\sum_{i=1}^{n} T_j(x_i);\ j = 1,\ldots,s\right\},$$

is a sufficient statistic and induces a standard exponential statistical space.

Proof The factorization theorem proves easily that T is sufficient. To compute the distribution of T perform the following transformations on the variables and on the parameters,

$$\mathbf{T}(x) = (T_1(x),\ldots,T_s(x)), \qquad x\in\mathbb{R}^k,$$

$$\boldsymbol{\theta} = (\theta_1,\ldots,\theta_s), \qquad \theta_j = Q_j(\theta),\quad j = 1,\ldots,s.$$

We call $\boldsymbol{\theta}$ the natural parameter. On the other hand, let m be the measure on $(\mathbb{R}^s, \mathscr{B}_{\mathbb{R}^s})$ which is the image of the measure v on $(\mathbb{R}^k, \mathscr{B}_{\mathbb{R}^k})$ defined by

$$dv = h\,d\mu,$$

induced by the mapping T, where μ is the measure which dominates the exponential statistical space. We have

$$T(x_1,\ldots,x_n) = \mathbf{T}(x_1) + \cdots + \mathbf{T}(x_n),$$

and the statistics $\mathbf{T}(x_1),\ldots,\mathbf{T}(x_n)$ are independent, each having the same probability density function,

$$\frac{\exp(\langle\boldsymbol{\theta}|\mathbf{x}\rangle)}{L_m(-\boldsymbol{\theta})}, \qquad \mathbf{x}\in\mathbb{R}^s.$$

Then, Theorem 2.4, implies that T has the probability density function

$$\frac{\exp(\langle\boldsymbol{\theta}|\mathbf{x}\rangle)}{[L_m(-\boldsymbol{\theta})]^n}. \qquad \blacksquare$$

Remark If $kn \leqslant s$, no reduction is possible. However, if $kn > s$, the dimension of the sufficient statistic T is s, where s is not a function of n. See Exercise 12.7.

Theorem 2 *For an exponential statistical space, let $\tilde{\Theta}$ be the subset of $\mathbb{R}^s$, defined by*

$$\boldsymbol{\theta} = (\theta_1, \ldots, \theta_s) \in \tilde{\Theta} \quad \Leftrightarrow \quad \{\theta_j = Q_j(\theta); j = 1, \ldots, s; \theta \in \Theta\}.$$

If $\dim \tilde{\Theta} = s$, *the statistic T of Theorem* 1 *is $\mathscr{P}$-minimum sufficient.*

Proof Let $p_\theta(x) = C(\theta) \exp\{\sum_{j=1}^{s} Q_j(\theta) T_j(x)\} h(x)$ and let $\mathscr{L}$ be the linear space which is generated by the functions $\log(p_\theta(x))/(p_{\theta_0}(x))$ when θ runs through Θ. The functions $1, T_1, \ldots, T_s$ generate $\mathscr{L}$ and we can apply Theorem 2.5.6 if we show that the functions T_j belong to $\mathscr{L}$. But $\tilde{\Theta}$ has dimension s and thus $\tilde{\Theta} - \boldsymbol{\theta}_0$ generates $\mathbb{R}^s$; i.e., there exist $\{\boldsymbol{\theta}_k \in \tilde{\Theta}\}$ and real numbers b_k (the index k has at most s values), such that

$$(1, 0, \ldots, 0) = \sum_k b_k(\boldsymbol{\theta}_k - \boldsymbol{\theta}_0).$$

Now using the definition of $\tilde{\Theta}$ we have

$$\sum_k b_k[Q_1(\theta^k) - Q_1(\theta^0)] = 1,$$

$$\sum_k b_k[Q_j(\theta^k) - Q_j(\theta^0)] = 0, \qquad j = 2, \ldots, s,$$

where θ^0 and θ^k belong to Θ. This implies that T_1 is a linear combination of functions $\log(p_\theta/p_{\theta_0})$ and thus belongs to $\mathscr{L}$. The same argument holds for $T_2, \ldots, T_s$ and the proof is completed. ∎

4. Incomplete Exponential Statistical Spaces

Let $(\Omega, \mathscr{A}, P_\theta, \theta \in \Theta)$ be a statistical space and let f_0 be an integrable statistic whose image is φ_0. We denote by Θ_0 the subset of Θ, defined by

$$\varphi_0(\theta) = 0, \qquad \theta \in \Theta.$$

It is obvious that the statistical space $(\Omega, \mathscr{A}; P_\theta, \theta \in \Theta_0)$ is not complete. More generally, we consider the case of exponential statistical spaces which are incomplete because there exist some constraints on the parameter (see Exercises 11 and 15).

The Behrens–Fisher problem and some analysis of variance (Model II) problems lead to such incomplete exponential statistical spaces. Determinating statistics having a zero image is useful for testing or estimating purposes, but is always difficult (see Section 12.6).

Definition 1 Assume that the following standard exponential statistical space

$$\left[\mathbb{R}^k, \mathscr{B}_{\mathbb{R}^k}; \frac{dP_\theta}{d\mu} = C(\theta)e^{\langle\theta|x\rangle}, \theta \in \Theta_0\right]$$

is not complete. If there exists a subset Θ of D' whose interior set is not empty and if there exist some real functions $\pi_1, \ldots, \pi_r$ $(r < k)$, on Θ, such that

$$\theta \in \Theta_0 \quad \Leftrightarrow \quad \theta \in \Theta, \qquad \pi_j(\theta) = 0, \quad \forall j = 1, \ldots, r,$$

then we say that this exponential statistical space is *analytically incomplete.*

The linear space of statistics having a zero image plays an important part in the study of an incomplete statistical space.

Theorem 1 *For the incomplete statistical space of Definition* 1, *assume that there exists a linear subspace W of $\mathbb{R}^k$ such that for every $\theta \in \mathbb{R}^k$, if $(W + \theta) \cap \Theta_0 \neq \varnothing$ then $(W + \theta) \cap \Theta_0$ has an interior nonempty set in $(W + \theta)$. Then a statistic has a zero image if and only if it has zero conditional expectation given the projection on W.*

Proof Separate every vector of $\mathbb{R}^k$ into its projections on W and on $W^\perp$, and write

$$x = v + u, \qquad \theta = \mu + \lambda.$$

Let Λ_0 be the projection of Θ_0 on $W^\perp$. Then for every λ belonging to Λ_0 the interior set in $\lambda + W$ of $\Theta_0 \cap (\lambda + W)$ is not empty. Performing a linear transformation on the parameters and on the variables, we can apply Theorem 2.3 to obtain

$$E_{P_\theta}(X) = \frac{1}{L_m(-\theta)} \int E_{P_\theta}(X \mid v) e^{\langle\mu|v\rangle} \, dm_\lambda(v).$$

Using part (ii) of that theorem, it can be seen that $E_{P_\theta}(X \mid v)$ does not depend on μ. Thus it can be denoted by $E_\lambda(X \mid v)$. Then if the statistic X has a zero image for every λ belonging to Λ_0,

$$\int E_\lambda(X \mid v) e^{\langle\mu|v\rangle} \, dm_\lambda(v) = 0, \qquad \forall \mu \in \Theta_0^\lambda,$$

where Θ_0^λ is the intersection of Θ_0 and $(\lambda + W)$. Since Θ_0^λ has a nonempty interior set in $(\lambda + W)$, Theorem 2.1 shows that $E_\lambda(X \mid v)$ is zero m_λ-almost everywhere. Conversely if $E_\lambda(X \mid v)$ is zero m_λ-almost everywhere it is obvious that X has zero expectation. ∎

Remark If Θ_0 is a cylinder, the assumptions of the theorem are met. This is the case if μ is a nuisance parameter and if the constraints are given only on λ.

Theorem 2 *Consider an incomplete statistical space, satisfying the assumptions of Theorem 1. If Z is a real-valued statistic whose square is integrable and if Z depends only on the projection on W, then Z is an unbiased estimate of its image having minimum variance.*

Proof Let X be a statistic which has a zero image. As in the proof of Theorem 1, since Z does not depend on v, we have

$$E_\lambda(XZ \mid v) = ZE_\lambda(X \mid v) = 0, \qquad \forall \lambda \in \Lambda_0.$$

By integration, we deduce that XZ has a zero image. The proof is completed by using Theorem 6.2.3. ∎

The converse of this theorem is true (see Linnik [42, p. 91]). However, it should be noted that under general conditions, one can have estimators having minimum variance only in the incomplete exponential statistical spaces where Θ_0 has the shape of a cylinder.

5. The Behrens–Fisher Problem

Consider the statistical space Σ:

$$[\mathbb{R}, \mathscr{B}_{\mathbb{R}}; N(m, \sigma^2), (m, \sigma) \in \mathbb{R} \times \mathbb{R}^+]^n \otimes [\mathbb{R}, \mathscr{B}_{\mathbb{R}}; N(m', \sigma'^2), (m', \sigma') \in \mathbb{R} \times \mathbb{R}^+]^{n'}.$$

Using obvious notation, the following set of statistics is a sufficient statistic:

$$x = \frac{1}{n}\sum_1^n x_i, \qquad y = \frac{1}{n'}\sum_1^{n'} y_i,$$

$$s = \frac{1}{n}\sum_1^n x_i^2 - x^2, \qquad t = \frac{1}{n'}\sum_1^{n'} y_i^2 - y^2.$$

These four statistics are independent and have distributions,

$$N\left(m, \frac{\sigma^2}{n}\right), N\left(m', \frac{\sigma'^2}{n'}\right), \Gamma\left(\frac{n-1}{2}, 2\sigma^2\right), \Gamma\left(\frac{n'-1}{2}, 2\sigma'^2\right).$$

Definition 1 The problem of testing the hypothesis $\{m = m'\}$ against $\{m \neq m'\}$ on the statistical space Σ is called the *Behrens–Fisher problem.*

Definition 2 When $n = n'$, any free test having the critical region,

$$(x - y)^2 \geqslant C\left(\sum_1^n (x_i - x - y_i + y)^2\right)$$

where C is a given constant, is called the *Bartlett–Scheffé test* for the Behrens–Fisher problem.

This test is the Student test of the differences,

$$z_i = x_i - y_i, \qquad i = 1, \ldots, n,$$

and then it is easy to compute the constant C corresponding to a given level of significance. For the definition of this test when $n \neq n'$, see Scheffé [61].

Under very reasonable assumptions (see, e.g. [42; p. 57]) we can restrict ourselves to homogeneous tests, i.e., to tests which are based on the pair,

$$\frac{(\bar{x} - \bar{y})^2}{s}, \frac{s}{t}.$$

Linnik has proved [42; p. 65] that, for every given level of significance, if $|n - n'|$ is not an even integer then there exists a homogeneous free deterministic test for the Behrens–Fisher problem. But he also proved [41] that the boundary of the critical region of this deterministic test is very irregular. In conclusion, we note that the difficulty of this problem is due to the incomplete statistical space corresponding to the null hypothesis (see Exercise 16).

Exercises

1. Compute the Laplace transform of the measure m on $(\mathbb{R}^+, \mathscr{B}_{\mathbb{R}^+})$, defined by

$$dm/dx = x^\alpha,$$

where α is a positive number. Determine the set D_m.

2. Let m be the measure on $(\mathbb{R}^3, \mathscr{B}_{\mathbb{R}^3})$, defined by

$$\frac{dm}{dx\,dy\,dz} = \begin{cases} (xy - z^2)^\alpha & \text{if } x \geqslant 0, \quad y \geqslant 0, \quad z^2 \leqslant xy, \\ 0 & \text{if not,} \end{cases}$$

where α is a given number. Find the values of α for which m is a smooth measure and in this case, compute the Laplace transform of m and determine the set D_m.

3. Find the smooth measure m on $(\mathbb{R}^s, \mathscr{B}_{\mathbb{R}^s})$ such that its Laplace transform is defined by

$$L_m(\theta_1, \ldots, \theta_s) = (a_1 e^{\theta_1} + \cdots + a_s e^{\theta_s})^n,$$

where $a_1, \ldots, a_s$ are given real numbers.

4. Show that a sample of size n from the multivariate normal distribution is an exponential statistical space and, for this special case, verify Theorem 3.1.

5. Find conditions on the subset $\Theta \in \mathbb{R}^2$ for which the statistical space,

$$[\mathbb{R}^+, \mathscr{B}_{\mathbb{R}^+}; \Gamma(a, \gamma, \lambda), (\gamma, \lambda) \in \Theta]$$

is an exponential one.

6. Find all the distributions of the exponential family for which the sample mean in a sample of size n is a sufficient statistic.

7. Show that the product or the weak product of two exponential statistical spaces is an exponential statistical space.

8. Prove Theorem 2.2.

9. Using Theorem 3.1, show that if the functions Q_j $(j = 1, \ldots, s)$ are linearly independent, then T is a $\mathscr{P}$-minimum sufficient statistic.

10. Consider Theorem 3.2 when $\dim \tilde{\Theta} < s$. Find a $\mathscr{P}$-minimum sufficient statistic.

11. Consider Theorem 9.9.1. Show that the statistics S_I, where I runs through the family of all nonempty subsets of K, induce a complete standard exponential statistical space. Compute the natural parameters. If some of the σ_J are zero, show that one has an analytically incomplete exponential statistical space.

12. Consider the complete standard exponential statistical space,

$$\left[\mathbb{R}^s, \mathscr{B}_{\mathbb{R}^s}, \frac{dP_\theta}{dx} = \frac{e^{\langle\theta|x\rangle}}{l_f(-\theta)} f(x), \theta \in \Theta\right],$$

where l_f is the Laplace transform of the function f (see Section 12.1) and let g be a real function of θ.

(a) Write the equation which is satisfied by the unbiased estimate of g, when it exists.

(b) In the case where f has derivatives of any order and g is a polynomial, compute the unbiased estimate of g and show that it has minimum variance.

(c) Find all functions of θ such that their unbiased estimates have maximum efficiency.

13. Consider a sample of size n from the normal distribution on $\mathbb{R}$ and let Φ be a function such that $[\Phi(x)] \leqslant e^{K|x|}$, where K is a given constant.

(a) Show that the function

$$f(m, \sigma) = \frac{1}{\sigma(2\pi)^{1/2}} \int_{-\infty}^{+\infty} \Phi(x) \exp[-(x - m)^2/2\sigma^2]\, dx$$

is defined for every m and σ.

(b) Without computations find an unbiased and convergent estimator of f.

(c) Find a lower bound for the variance of any unbiased estimator of f.

(d) Find the unbiased estimator of f having minimum variance. Compute it, using only an integral with respect to one variable.

14. Consider the statistical space Σ corresponding to the Behrens–Fisher problem (Section 5). Find a sufficient $\mathscr{P}$-minimum statistic on $(\mathbb{R}^4, \mathscr{B}_{\mathbb{R}^4})$, which induces a standard exponential statistical space. Find a statistic on $(\mathbb{R}^3, \mathscr{B}_{\mathbb{R}^3})$ which has the same properties, when the ratio of the variances is known.

15. Consider the Behrens–Fisher problem and the statistical space which is induced by the sufficient statistic. Assume that the null hypothesis is true.

(a) Show that this statistical space is an analytic incomplete exponential statistical space.

(b) Determine the statistical space induced by $\{(x - y)^2, s, t\}$ (notation of Section 5), and find a sufficient statistic for this statistical space.

16. Consider the statistical space,

$$[\mathbb{R}, \mathscr{B}_{\mathbb{R}}; N(m, \sigma^2), (m, \sigma^2) \in \Theta \subset \mathbb{R} \times \mathbb{R}^+]^n.$$

(a) Find conditions on n and Θ under which there exists a complete $\mathscr{P}$-minimum sufficient statistic which has values on $(\mathbb{R}^2, \mathscr{B}_{\mathbb{R}^2})$. Compute this statistic and the statistical space that it induces.

(b) Study the special case where Θ is a linear subspace of $\mathbb{R}^2$.

17. Consider the statistical space,

$$[\mathbb{R}^k, \mathscr{B}_{\mathbb{R}^k}; N(m, \Lambda), (m, \Lambda) \in \Theta \subset \mathbb{R}^k \times \mathscr{S}_k^+]^n.$$

Find a complete and $\mathscr{P}$-minimum sufficient statistic T. In the case where T has dimension $\frac{1}{2}(k(k + 3))$, $n > k$, compute its distribution.

18. Compute the information and the maximum likelihood statistic associated with a standard exponential statistical space.

CHAPTER 11

Testing Hypotheses on Exponential Statistical Spaces

This chapter shows that, for exponential spaces, if there is only one unknown scalar parameter, then optimal tests exist and are very simple. Moreover, if there is one principal scalar parameter, the others being nuisance parameters, the optimal tests are also simple. The results are due essentially to Lehmann [44].

1. The Case of One Unknown Scalar Parameter

Consider an exponential statistical space with $\dim \tilde{\Theta} = 1$. There exists a sufficient statistic which induces a standard exponential statistical space defined on $\mathbb{R}$. Then, any problem of testing a hypothesis on the original statistical space can be reduced to a problem of testing a hypothesis on the statistical space

$$\{\mathbb{R}, \mathscr{B}_{\mathbb{R}}; dP_\theta(t) = Q(\theta)e^{\theta t}\, d\nu(t), \theta \in D'_\nu\}, \tag{1}$$

where the notation is that of Section 10.2, ν is a smooth positive measure on $(\mathbb{R}, \mathscr{B}_{\mathbb{R}})$, D'_ν is the symmetric set of D_ν with respect to the origin, and

$$Q(\theta) = \frac{1}{L_\nu(-\theta)}.$$

As already noted in the remark following Definition 10.1.2, the set D_ν is an interval.

We now consider the statistical space (1) and assume that ν is not a Dirac measure because this case is of no interest. Using Theorem 10.2.1 we prove that if the images β_Φ and $\beta_{\Phi'}$ of two tests Φ and Φ' are equal on an open set of D'_ν, then these two tests are equal ν-almost everywhere.

Lemma 1 *The image β_Φ of every test Φ is an analytic function in the interior of D'_ν.*

Proof Let Φ be a test and ν' be the positive measure defined by

$$d\nu' = \Phi \, d\nu.$$

The condition

$$0 \leqslant \Phi \leqslant 1$$

implies that

$$D_{\nu'} \supset D_\nu.$$

Then

$$D^0_{\nu'} \supset D^0_\nu,$$

where D^0_ν is the interior of D_ν. Moreover, we have

$$\beta_\Phi(\theta) = \frac{L_{\nu'}(-\theta)}{L_\nu(-\theta)}.$$

The functions $L'_{\nu'}$ and L_ν are positive and, using Theorem 10.1.2, analytic in D^0_ν. ∎

We can compute the derivative of β_Φ,

$$\theta \in D'^0_\nu, \qquad \frac{d}{d\theta}\beta_\Phi(\theta) = \int_{\mathbb{R}} \Phi[Q'(\theta) + tQ(\theta)]e^{\theta t}\, d\nu(t),$$

and by using Theorem 10.2.2, we have

$$Q'(\theta) = -Q^2(\theta) \int_{\mathbb{R}} te^{\theta t}\, d\nu(t).$$

Then

$$\theta \in D'^0_\nu, \qquad \frac{d}{d\theta}\beta_\Phi(\theta) = E_\theta[t\Phi(t)] - E_\theta(t)E_\theta(\Phi(t)). \tag{2}$$

∎

Lemma 2 *Let Φ be a test on the statistical space* (1). *If there exists a value of θ such that*

$$\beta_\Phi(\theta) = 1 \qquad \text{(or 0)},$$

then

$$\Phi = 1 \qquad \text{(or 0)}, \quad \nu\text{-a.e.}$$

Proof The function

$$p_\theta(t) = \frac{dP_\theta}{d\nu} = Q(\theta)e^{\theta t}$$

is positive. If, for some nonnegative function f, there exists a specific value of θ for which

$$\int f\, dP_\theta = 0,$$

then

$$f = 0, \qquad \nu\text{-a.e.}$$

∎

2. One-Sided and Two-Sided Tests on a One-Dimensional Exponential Statistical Space

Definition 1 We call a *right-sided* test (or *left-sided* test) any test Φ in which

(i) $\forall\theta \in D'_\nu, \qquad 0 < \beta_\Phi(\theta) < 1,$

(ii) there exists a constant C such that

$$t > C \;\Rightarrow\; \Phi = 1 \qquad \text{(or 0)},$$

$$t < C \;\Rightarrow\; \Phi = 0 \qquad \text{(or 1)}.$$

Using Lemma 1.2, we can prove that if the test Φ is one-sided, then Φ *and* $1 - \Phi$ cannot be zero ν-a.e. Then one of the following conditions must be satisfied.

(1) $0 < \Phi(C) < 1, \qquad \nu[\{C\}] > 0 \quad \text{or} \quad \nu((-\infty, C[)\nu(]C, +\infty)) > 0;$

(2) $\Phi(C) = 0, \qquad \nu(-\infty, C]) > 0, \quad \nu(]C, +\infty) > 0;$

(3) $\Phi(C) = 1, \qquad \nu(-\infty, C[) > 0, \quad \nu([C, +\infty) > 0.$

Thus a right-sided test always accepts when the statistic is less than some number, always rejects when the statistic is greater than that same number and may be randomized when the statistic equals that number. Note that a left-sided test corresponds to a right-sided test by the transformation $\Phi \to 1 - \Phi$.

Lemma 1 *For given points θ and θ' in D'_v $(\theta < \theta')$, any right-sided test (or left-sided test) is an admissible test of the hypothesis $\{\theta\}$ against the hypothesis $\{\theta'\}$ (or $\{\theta'\}$ against $\{\theta\}$).*

Proof Let Φ be a right-sided test and let C be the corresponding constant (cf. Definition 1). If we denote

$$A = \frac{Q(\theta')}{Q(\theta)} e^{(\theta' - \theta)C} > 0,$$

then we have

$$p_{\theta'} > Ap_\theta \quad \Rightarrow \quad \Phi = 1, \qquad p_{\theta'} < Ap_\theta \quad \Rightarrow \quad \Phi = 0.$$

Since A is positive, Theorem 5.5.3 shows that Φ is an admissible test.

Corollary 1 *The image of every right-sided test (or left-sided test) is a strictly increasing (or decreasing) function.*

Proof Let Φ be a right-sided test; for every $\theta < \theta'$ we have

$$P_\theta \neq P_{\theta'}.$$

Then, using Lemma 1, we have

$$\beta_\Phi(\theta) \leqslant \beta_\Phi(\theta')$$

since Φ is better than the trivial test which has the same level of significance. Now we prove, by *reductio ad absurdum*, that one cannot have

$$\beta_\Phi(\theta) = \beta_\Phi(\theta'),$$

since Φ is an admissible test, for which

$$0 < \beta_\Phi(\theta) < 1.$$

If the previous equality were satisfied, the locus $\mathscr{R}$ of pairs $[\beta_\psi(\theta), \beta_\psi(\theta')]$, where ψ runs through all possible tests, would reduce to a 45-degree line and then P_θ and $P_{\theta'}$ would be equal. By hypothesis this is impossible. ∎

Corollary 2 *For every $\theta \in D'_\nu$ and $\alpha \in]0, 1[$, there exists a right-sided test Φ' and a left-sided test Φ for which*

$$\beta_\Phi(\theta) = \beta_{\Phi'}(\theta) = \alpha.$$

These two tests are unique (ν-almost everywhere), and we call them one-sided *tests having level of significance α at the point θ.*

Proof Let $F_\theta(t)$ be the cumulative distribution function of P_θ and Φ be a left-sided test. We have

$$\beta_\Phi(\theta) = F_\theta(C) + \Phi(C)[F_\theta(C+0) - F_\theta(C)].$$

Then by an argument analogous to that of Section 5.5, we prove that

(a) if there exists a constant C such that

$$F_\theta(C) = F_\theta(C+0) = \alpha,$$

then denote by C_1 the smallest value of C such that

$$F_\theta(C+0) \geqslant \alpha$$

and by C_2 the largest value of C such that

$$F_\theta(C) \leqslant \alpha.$$

Then all the left-sided tests satisfying the conditions of the corollary are defined by choosing C in $[C_1, C_2]$ and $\Phi(C)$ such that

$$[1 - \Phi(C_1)]\nu(\{C_1\}) = \Phi(C_2)\nu(\{C_2\}) = 0.$$

Since it is obvious that $\nu(]C_1, C_2[) = 0$, these tests are the same (ν-a.e.).

(b) if Assumption (a) is not satisfied, there exists a unique value C such that

$$F_\theta(C) \leqslant \alpha \leqslant F_\theta(C+0),$$

and the test Φ is defined by

$$\Phi = \begin{cases} 1 & \text{if } t < C, \\ 0 & \text{if } t > C, \\ \dfrac{\alpha - F_\theta(C)}{F_\theta(C+0) - F_\theta(C)} & \text{if } t = C. \end{cases}$$

In both cases, since α is different from 0 and 1, the condition

$$0 < \beta_\Phi(\theta) < 1$$

is satisfied; otherwise, using Lemma 1.2, we would have

$$\beta_\Phi = 0 \qquad \text{or} \qquad \beta_\Phi = 1. \qquad \blacksquare$$

Definition 2 A test Φ is called *two-sided* if

(i) Φ is not equal (ν-a.e.) to 0,
(ii) Φ is not equal (ν-a.e.) to 1,
(iii) Φ is not equal (ν-a.e.) to some one-sided test,
(iv) there exist two constants C and C' such that

$$C < t < C' \quad \Rightarrow \quad \Phi = 1,$$

$$t < C \qquad \text{or} \qquad t > C' \quad \Rightarrow \quad \Phi = 0.$$

Part (iii) is necessary to assure that the acceptance region consists of two parts. For given C, C', $\Phi(C)$, $\Phi(C')$ the practical verification of part (iii) consists in showing that the proposed test is ν-equivalent neither to the right-sided test defined by C, $\Phi(C)$, nor to the left-sided test defined by C', $\Phi(C')$.

Lemma 2 *For all triples, θ, θ_1, θ', belonging to D'_ν ($\theta < \theta_1 < \theta'$) every two-sided test is an admissible test of the hypothesis $\{\theta\} \cup \{\theta'\}$ against θ_1.*

Proof Let Φ be a two-sided test and let C and C' be the corresponding constants. It is easy to verify that the roots A and A' of the linear system,

$$AQ(\theta)e^{\theta C} + A'Q(\theta')e^{\theta' C} = Q(\theta_1)e^{\theta_1 C}; \qquad AQ(\theta)e^{\theta C'} + A'Q(\theta')e^{\theta' C'} = Q(\theta_1)e^{\theta_1 C'}$$

are positive. Then we have

$$p_{\theta_1} > Ap_\theta + A'p_{\theta'} \quad \Rightarrow \quad \Phi = 1; \qquad p_{\theta_1} < Ap_\theta + A'p_{\theta'} \quad \Rightarrow \quad \Phi = 0, \quad (1)$$

because inequality (1) is equivalent to

$$Q(\theta_1)e^{\theta_1 t} > AQ(\theta)e^{\theta t} + A'Q(\theta')e^{\theta' t},$$

and thus defines a bounded interval, (C, C'). Then, using Theorem 5.5.3, we prove that Φ is admissible. $\blacksquare$

Corollary 3 *The image of any two-sided test is either strictly monotone, or it is strictly increasing on the left of a suitable point θ^* of D'_ν and strictly decreasing on the right of this point.*

Proof Using Lemma 2, we see that for all $\theta < \theta_1 < \theta' \in D'_\nu$,

$$\beta_\Phi(\theta_1) \geqslant \inf[\beta_\Phi(\theta), \beta_\Phi(\theta')].$$

We shall prove, by *reductio ad absurdum*, that we cannot have

$$\beta_\Phi(\theta) = \beta_\Phi(\theta') = \beta_\Phi(\theta_1).$$

Let $\mathscr{R}$ be the locus in $\mathbb{R}^3$ of the points

$$\boldsymbol{\beta}_{\Phi'} = [\beta_{\Phi'}(\theta), \beta_{\Phi'}(\theta'), \beta_{\Phi'}(\theta_1)]$$

where Φ' runs through all possible tests. We know that $\mathscr{R}$ is not reduced to its diagonal and, since Φ is an admissible test, $\boldsymbol{\beta}_\Phi$ is a boundary point of $\mathscr{R}$. Then, if the previous equations are satisfied, $\mathscr{R}$ has dimension 2 and thus one has

$$\lambda_1 P_{\theta_1} + \lambda P_\theta + \lambda' P_{\theta_1} = 0.$$

Finally, it follows that in this case no two-sided test exists. ∎

Lemma 3 *If ν is not concentrated on two points, for all pairs $\theta \in D'_\nu$, $\theta' \in D'_\nu$, $\theta < \theta'$ and $\alpha \in]0, 1[$, there exists a unique (ν-a.e.) two-sided test such that*

$$\beta_\Phi(\theta) = \beta_\Phi(\theta') = \alpha. \tag{2}$$

We call this test the two-sided test having level of significance α at the points θ and θ'. Moreover there exists a point θ^* of $]\theta, \theta'[$ such that β_Φ is strictly increasing on the left of θ^* and strictly decreasing on the right of θ^*. For any other test ψ satisfying (2),

$$\begin{cases} \beta_\Phi(\theta_1) < \beta_\psi(\theta_1) & \text{if } \theta_1 \in D'_\nu - [\theta, \theta'] \\ \beta_\psi(\theta_1) < \beta_\Phi(\theta_1) & \text{if } \theta_1 \in]\theta, \theta'[. \end{cases}$$

Proof Let θ_1 be a point of D'_ν other than θ and θ' and let $\mathscr{R}$ be the locus considered in the proof of Corollary 3. The set $\mathscr{R}$ is convex and closed, and its intersection with the straight line defined by (2) is an interval. Since ν is not concentrated on two points, this interval does not reduce to one point; denote its endpoints by $\boldsymbol{\beta}_\Phi$ and $\boldsymbol{\beta}_{\Phi'}$, $\boldsymbol{\beta}_\Phi \neq \boldsymbol{\beta}_{\Phi'}$. Using Theorem 5.4.1 we can prove that there exist constants A, A', B, B', C, C' such that, ν-a.e.,

$$Ap_{\theta_1} > Bp_\theta + Cp_{\theta'} \;\Rightarrow\; \Phi = 1, \qquad A'p_{\theta_1} > B'p_\theta + C'p_{\theta'} \;\Rightarrow\; \Phi' = 1,$$

$$Ap_{\theta_1} < Bp_\theta + Cp_{\theta'} \;\Rightarrow\; \Phi = 0, \qquad A'p_{\theta_1} < B'p_\theta + C'p_{\theta'} \;\Rightarrow\; \Phi' = 0.$$

But it is easy to verify that the first inequality is equivalent to

$$AQ(\theta_1) > BQ(\theta)e^{(\theta - \theta_1)t} + CQ(\theta')e^{(\theta' - \theta_1)t},$$

which always defines either an interval or the complementary set of an interval. This argument holds for each of the inequalities.

On the other hand, using Corollary 1, we see that a one-sided test cannot satisfy (2). Then either Φ or $1 - \Phi$ is a two sided test and either Φ' or $1 - \Phi'$ is a two-sided test. Note that α is an interior point of $[\beta_\Phi(\theta_1), \beta_{\Phi'}(\theta_1)]$, and, without loss of generality, we can assume that

$$\beta_\Phi(\theta_1) > \alpha > \beta_{\Phi'}(\theta_1).$$

Using Corollary 3, we see that if θ_1 belongs to $]\theta, \theta'[$, then Φ and $1 - \Phi'$ are both two-sided tests; if θ_1 does not belong to $]\theta, \theta'[$, then Φ' and $1 - \Phi$ are both two-sided tests. Thus, in both cases we have found a two-sided test which satisfies (2). Since Condition (2) is satisfied, the image of this test cannot be strictly monotone and thus the first part of the proof is completed.

Now, consider a two-sided test Φ which satisfies (2). Using Lemma 2, it is obvious that we have

$$\beta_\Phi(\theta) = \begin{cases} \inf\limits_{\psi} \beta_\psi(\theta) & \text{if} \quad \theta \notin [\theta, \theta'], \\ \sup\limits_{\psi} \beta_\psi(\theta) & \text{if} \quad \theta \in [\theta, \theta'], \end{cases}$$

where ψ runs through all possible tests satisfying (2). Hence any two-sided test which satisfies (2) has the same image as Φ. Since the statistical space is complete, Φ is unique ν-a.e.

Finally, using Lemma 2 we see that, if there exists a θ_1 for which

$$\beta_\Phi(\theta_1) = \beta_\psi(\theta_1),$$

then this equation is true for every θ_1. ∎

Lemma 4 *If the measure ν is not concentrated on two points, then for every θ belonging to the interior set of D'_ν and every α belonging to $]0, 1[$, there exists a unique (ν-a.e.) two-sided test such that*

$$\beta_\Phi(\theta) = \alpha, \qquad \frac{d}{d\theta}\beta_\Phi(\theta) = 0. \tag{3}$$

This test is called the two-sided test having level of significance α at the point θ. The image of β_Φ is an increasing function on the left of θ and decreasing function on the right of θ. It is the lower bound of the images of all the tests which satisfy (3).

Proof The proof is analogous to that of Lemma 3. Using (1.2) we see that (3) is equivalent to

$$\beta_\Phi(\theta) = \alpha, \qquad \beta_{t\Phi}(\theta) = \alpha\beta_t(\theta). \tag{3'}$$

Let θ' be a point of D'_ν which is different from θ; let $\mathscr{R}'$ be the closed convex set of $\mathbb{R}^3$, defined by the points

$$\boldsymbol{\beta}'_{\Phi} = [\beta_{\Phi}(\theta), \beta_{t\Phi}(\theta), \beta_{\Phi}(\theta')],$$

where Φ runs through all possible tests. The intersection of $\mathscr{R}'$ with the straight line defined by (3′) is a closed interval $(\boldsymbol{\beta}'_{\Phi}, \boldsymbol{\beta}'_{\Phi'})$ which is not reduced to a point and satisfies

$$\beta_{\Phi}(\theta') \leqslant \alpha \leqslant \beta_{\Phi'}(\theta'). \tag{4}$$

Using Theorem 5.4.1 there exist constants A, A', B, B', C, C' such that

$$\begin{cases} Ap_{\theta'} > Bp_{\theta} + Ctp_{\theta} & \Rightarrow \quad \Phi = 1, \qquad \nu\text{-a.e.}, \\ Ap_{\theta'} < Bp_{\theta} + Ctp_{\theta} & \Rightarrow \quad \Phi = 0, \qquad \nu\text{-a.e.}, \end{cases}$$
$$\begin{cases} A'p_{\theta'} > B'p_{\theta} + C'tp_{\theta} & \Rightarrow \quad \Phi' = 1, \qquad \nu\text{-a.e.}, \\ A'p_{\theta'} < B'p_{\theta} + C'tp_{\theta} & \Rightarrow \quad \Phi' = 0, \qquad \nu\text{-a.e.} \end{cases} \tag{5}$$

The first of these inequalities is equivalent to

$$A\frac{Q(\theta')}{Q(\theta)}e^{(\theta'-\theta)t} > B + B't,$$

and always defines either an interval or its complement. This argument holds for each of the inequalities. However, using Corollary 1, we see that a one-sided test cannot satisfy Condition (3′). Thus Corollary 3 and Inequalities (4) show that Φ and $1 - \Phi'$ are both two-sided tests and that Inequalities (4) are, in fact, strict inequalities. Moreover, (3) implies that the image of Φ cannot be strictly monotonic, and thus it is increasing on the left of θ and decreasing on the right of θ.

Now, let Φ be a two-sided test; it is easy to prove that for every θ' there exist constants A, B, C such that Conditions (5) are satisfied. Then, if Φ satisfies (3), $\beta_{\Phi}(\theta')$ is the lower endpoint of the interval (4). This argument shows that the image of Φ is the lower endpoint of the images of all tests, satisfying (3). Thus, Φ is unique (ν-a.e.). ∎

To compute the two-sided test having level of significance α at points θ, θ', $\theta < \theta'$, we must solve the following system

$$0 \leqslant \gamma, \qquad \gamma' \leqslant 1,$$

$$F_{\theta}(C') - F_{\theta}(C+0) + \gamma[F_{\theta}(C+0) - F_{\theta}(C)] + \gamma'[F_{\theta}(C'+0) - F_{\theta}(C')] = \alpha, \tag{6}$$

$$F_{\theta'}(C') - F_{\theta'}(C+0) + \gamma[F_{\theta'}(C+0) - F_{\theta'}(C)] + \gamma'[F_{\theta'}(C'+0) - F_{\theta'}(C')] = \alpha. \tag{7}$$

To compute the two-sided test having level of significance α at the point θ, we would have to replace Eq. (7) by an equation denoted as (8) which would be obtained by requiring that the power function have a zero derivative at θ. Equation (8) is not given. Note that a two-sided test is the difference between two left-sided tests, and introduce the following auxiliary variable y,

$$1 \geqslant y \geqslant \alpha, \qquad F_\theta(C') + \gamma'[F_\theta(C'+0) - F_\theta(C')] = y. \tag{9}$$

Equation (6) then becomes

$$F_\theta(C) + (1-\gamma)[F_\theta(C+0) - F_\theta(C)] = y - \alpha. \tag{10}$$

In order to continue the computation of two-sided tests we need:

Lemma 5 *Let θ be an interior point of D_v'. Let $\tilde{\beta}_w(\tau)$ be the common image of all the left-sided tests having level of significance w at the point θ. Let $K(w)$ and $H(w)$ be the functions defined by*

$$\forall w \in [0, 1], \qquad K(w) = \left.\frac{d\tilde{\beta}_w}{d\tau}\right|_{\tau=\theta}, \qquad H(w) = \tilde{\beta}_w(\theta'),$$

where θ' is a given point of D_v' larger than θ. Then the functions $K(w)$ and $H(w)$ are continuous and strictly convex and we have

$$K(0) = K(1) = 0, \quad H(0) = 0, \quad H(1) = 1,$$

$$\forall w \in \,]0, 1[, \qquad K(w) < 0, \quad w > H(w) > 0.$$

Proof It can easily be shown (Exercise 18) that for every w the function $\tilde{\beta}_w(\tau)$ has the following property:

(P1) *If Φ is a test having image $\beta_\Phi(\tau)$ and if there exists τ_0 such that $\beta_\Phi(\tau_0) = \tilde{\beta}_w(\tau_0)$, then*

$$\beta_\Phi(\tau) < \tilde{\beta}_w(\tau), \qquad \tau < \tau_0,$$

$$\beta_\Phi(\tau) > \tilde{\beta}_w(\tau), \qquad \tau > \tau_0.$$

Now let $\Phi_{w'}$ and $\Phi_{w''}$ be two left-sided tests having levels of significance w' and w'' respectively, at the point θ. Then the test $\Phi = (\Phi_{w'} + \Phi_{w''})/2$ has the image

$$\beta_\Phi(\tau) = \tfrac{1}{2}[\tilde{\beta}_{w'}(\tau) + \tilde{\beta}_{w''}(\tau)]$$

and then

$$\beta_\Phi(\theta) = \frac{w' + w''}{2}, \qquad \beta_\Phi(\theta') = \frac{H(w') + H(w'')}{2},$$

$$\left.\frac{d}{d\tau}\beta_\Phi(\tau)\right|_{\tau=\theta} = \frac{K(w') + K(w'')}{2}.$$

The proof is completed by applying Property P1 with

$$w = (w' + w'')/2. \quad ∎$$

Now we can find the two-sided tests satisfying (2). Equation (7) is equivalent to

$$H(y) - H(y - \alpha) = \alpha, \qquad 1 \geqslant y \geqslant \alpha. \tag{11}$$

The corresponding Eq. (8) (which was never actually displayed) in the case of a two-sided test having level of significance α at the point θ, is equivalent to

$$K(y) - K(y - \alpha) = 0, \qquad 1 \geqslant y \geqslant \alpha. \tag{12}$$

Using Lemma 5, we can prove that each of Eqs. (11) and (12) has a unique root, i.e. y. Now, y is given and then (9) and (10) determine C', γ', C, γ, in the same way as in Corollary 2. ∎

3. Optimal Tests on a One-Dimensional Exponential Statistical Space

Theorem 1 *Let Θ_0 be a nonempty subset of D'_v; let θ_1 be a point of $D'_v - \Theta_0$; let $\alpha \in]0, 1[$. Then there exists a test Φ which is strictly u.m.p. as a test of Θ_0 against θ_1 and has level of significance α. This test is defined by the following:*

(a) *If*

$$\Theta_0 \subset (-\infty, \theta_1[, \qquad (\text{or} \quad \Theta_0 \subset]\theta_1, +\infty)),$$

we designate by θ_0,

$$\theta_0 = \sup\{\theta | \theta \in \Theta_0\}, \qquad (\text{or} \quad \inf(\theta | \theta \in \Theta_0)).$$

Then Φ is the right-sided (or left-sided) test having level of significance α at the point θ_0

(b) *If the assumptions of* (a) *are not satisfied we let*

$$\theta'_0 = \sup\{\theta | \theta \in \Theta_0, \theta \leqslant \theta_1\},$$
$$\theta''_0 = \inf\{\theta | \theta \in \Theta_0, \theta \geqslant \theta_1\}.$$

If $\theta'_0 \neq \theta''_0$, then Φ is the two-sided test having level of significance α at the points θ'_0, θ''_0. If $\theta'_0 = \theta''_0$ then Φ is the two-sided test having level of significance α at the point θ_1.

Proof (a) Since Φ is obviously admissible, Theorem 5.4.3 can be used to imply (a).

(b) For every $\theta' \in \Theta_0$ and $\theta'' \in \Theta_0$, where

$$\theta' < \theta_0', \qquad \theta'' > \theta_0'',$$

Lemma 2.2 implies that Φ is an admissible test of the hypothesis $\{\theta'\} \cup \{\theta''\}$ against θ_1. Then Φ is an admissible test of Θ_0 against θ_1. If θ_1 is equal to either θ_0' or θ_0'', then every test ψ having level of significance α is such that $\beta_\psi(\theta_1) \leqslant \alpha$ and thus Φ is u.m.p. On the other hand, if

$$\theta_0' < \theta_1 < \theta_0'',$$

Lemma 2.2 implies that Φ is a u.m.p. test of $\{\theta_0'\} \cup \{\theta_0''\}$ against θ_1. Using Lemma 2.3, we see that Φ has level of significance α as a test of $\bar{\Theta}_0$ against θ_1, where $\bar{\Theta}_0$ is the closure of Θ_0. Then it is a u.m.p. test (Theorem 5.3.1). ∎

As a straightforward consequence of this theorem we have:

Corollary 1 *Let Θ_0 and Θ_1 be two disjoint hypotheses in D_v'; there exists a u.m.p. test of Θ_0 against Θ_1 if and only if Θ_0 and the convex hull of Θ_1 are disjoint.*

Theorem 2 *Let Θ_0 and Θ_1 be two disjoint hypotheses, such that there exist two points θ_0', θ_0'' (or a point θ_0), where*

$$\Theta_0 \subset [\theta_0', \theta_0''], \qquad \Theta_1 \subset \complement]\theta_0', \theta_0''[, \qquad \bar{\Theta}_1 \cap \bar{\Theta}_0 = \{\theta_0', \theta_0''\},$$

(or $\Theta_0 = \{\theta_0\}$, $\theta_0 \in \bar{\Theta}_1$ and θ_0 belongs to the interior set of the convex hull of Θ_1). Then the test $1 - \Phi$, where Φ is the two-sided test having the level of significance $1 - \alpha$ at the points θ_0' and θ_0'' (or θ_0), is a strictly u.m.p.u. test of Θ_0 against Θ_1 having level of significance α.

Proof Every unbiased test ψ of Θ_0 against Θ_1 satisfies

$$\beta_\psi(\theta_0') = \beta_\psi(\theta_0'') = \alpha \qquad \left(\text{or } \beta_\psi(\theta_0) = \alpha, \ \frac{d}{d\theta}\beta_\psi(\theta_0) = 0\right). \tag{1}$$

Then, Lemma 2.3 (or 2.4) shows that Φ is an unbiased test having level of significance α and that Φ is strictly better than other test ψ, satisfying (1). ∎

4. Testing Hypotheses in the Case of Nuisance Parameters

By the same reasoning as indicated in Section 1, it can be shown that every problem of testing hypotheses on an exponential statistical space can be reduced to a problem of testing hypotheses on the statistical space

$$[\mathbb{R}^s, \mathscr{B}^s; dP_\lambda = C(\lambda)e^{\langle\lambda|x\rangle}\, d\nu(x), \lambda \in D'_\nu],$$

where ν is a positive smooth measure on $(\mathbb{R}^s, \mathscr{B}_{\mathbb{R}^s})$ and

$$C(\lambda) = \frac{1}{L_\nu(-\lambda)}.$$

We now consider the case where the first component of λ is the principal parameter, the other components being nuisance parameters. We can write

$$x = (t, u), \qquad t \in \mathbb{R}, \quad u \in \mathbb{R}^{s-1}$$

$$\lambda = (\theta, \mu), \qquad \theta \in \mathbb{R}, \quad \mu \in \mathbb{R}^{s-1}, \quad (\theta, \mu) \in D'_\nu,$$

and denote by D^1_ν the projection of D'_ν on $\mathbb{R}$. Note that D^1_ν is an interval. We designate by ν_u the conditional measure given u, which corresponds to the measure q_v of Theorem 10.2.3.

Lemma 1 *For every $\theta \in D^1_\nu$ and $\alpha \in]0, 1[$, there exists a test $\Phi(t, u)$ such that for every $u \in \mathbb{R}^{s-1}$ the function*

$$t \to \Phi(t, u)$$

is a right-sided (left-sided) test defined on the standard exponential statistical space associated with the measure ν_u on $(\mathbb{R}, \mathscr{B}_{\mathbb{R}})$ and has level of significance α at the point θ. This test is called the right-sided (left-sided) conditional *test having level of significance α at the point θ.*

Proof We prove this lemma for the case of a right-sided test, defined by

$$\Phi(t, u) = \begin{cases} 1 & \text{if} \quad t < C(u), \\ \gamma(u) & \text{if} \quad t = C(u), \\ 0 & \text{if} \quad t > C(u). \end{cases} \tag{1}$$

For every given u, Corollary 2.2 shows that there exist $\gamma(u)$ and $C(u)$, such that the function $t \to \Phi(t, u)$ is a one-sided test having level of significance α at the point θ. Then, it is necessary only to find a measurable determination of the functions $\gamma(u)$ and $C(u)$, since in this case the function $(t, u) \to \Phi(t, u)$ is a measurable one. Let $F^u_\theta(t)$ be the conditional cumulative distribution function given u,

$$F^u_\theta(t) = \frac{1}{L_{\nu_u}(-\theta)} \int_{-\infty}^{t} e^{\theta t}\, d\nu_u(t),$$

and let (T, U) be a random vector having the distribution P_λ. Then $F^u_\theta(t)$ is a determination of $P_\lambda(T < t|U)$ and $F^u_\theta(t)$ is a measurable function of u, for every given t. On the other hand, for every given u, $F_\theta^u(t)$ is a cumulative distribution function. These two properties imply that $F_\theta^u(t)$ and $F_\theta^u(t+0)$ are measurable functions of the pair (t, u) (see Exercise 5). Thus the function

$$C(u) = \sup\{y \mid F_\theta^u(y) \leqslant \alpha\},$$

is measurable since

$$C(u) \leqslant x \quad \Leftrightarrow \quad F_\theta^u(x) \geqslant \alpha.$$

Now we can define the function $\gamma(u)$ by

$$\gamma(u) = \begin{cases} 0 & \text{if} \quad F_\theta^u(C(u)+0) = F_\theta^u(C(u)), \\ \dfrac{\alpha - F_\theta^u(C(u))}{F_\theta^u(C(u)+0) - F_\theta^u(C(u))} & \text{if not.} \end{cases}$$

The measurability of $C(u)$ and that of $F_\theta^u(t)$ as a function of (t, u) imply the measurability of $\gamma(u)$. ∎

Lemma 2 *For every $\theta \in D_\nu^1$ and every $\alpha \in \,]0, 1[$, there exists a test $\Phi(t, u)$ such that for every given $u \in \mathbb{R}^{s-1}$, the function $t \to \Phi(t, u)$ is a two-sided test defined on the standard exponential statistical space associated with the measure ν_u on $(\mathbb{R}, \mathscr{B}_\mathbb{R})$ and has level of significance α at the point θ. This test is called the* two-sided conditional *test having level of significance* α at the point θ.

Lemma 3 *For every $\theta \in D_\nu^1$ and $\theta' \in D_\nu^1$ and every $\alpha \in \,]0, 1[$ there exists a test $\Phi(t, u)$ such that for every given $u \in \mathbb{R}^{s-1}$ the function Φ, as function of t, is a two-sided test defined on the standard exponential statistical space associated with the measure ν_u on $(\mathbb{R}, \mathscr{B}_\mathbb{R})$ and has level of significance α at the points θ and θ'. This test is called a* two-sided conditional *test having level of significance* α at the points (θ, θ').

We prove these two lemmas together. We must find measurable functions $C(u)$, $C'(u)$, $\gamma(u)$, $\gamma'(u)$ such that, for every u, the test

$$t \to \Phi(t, u) = \begin{cases} 1 & \text{if} \quad t \in [C(u), C'(u)], \\ \gamma(u) & \text{if} \quad t = C(u), \\ \gamma'(u) & \text{if} \quad t = C'(u), \\ 0 & \text{otherwise.} \end{cases} \tag{2}$$

has level of significance α at the points θ, θ' (or at the point θ). Using Lemmas 2.3 and 2.4 we see that this is a problem of choice of a measurable determination. Consider Eqs. (2.6)–(2.10) and replace F_θ by F_θ^u. Then the functions

$H(w)$ and $K(w)$ become $H_u(w)$ and $K_u(w)$, respectively. The function

$$C(u, w) = \sup\{y \mid F_\theta^u(y) \leqslant w\}$$

has the following properties:

(i) For every w the function $u \to C(u, w)$ is measurable, as can be shown by using Lemma 1.

(ii) For every u, the function $w \to C(u, w)$ is an increasing function.

Then, C is a measurable function of (u, w), and $H_u(w)$ and $K_u(w)$ are measurable functions of (u, w) (see Exercise 9). Moreover, for every given u, the functions $K_u(w)$ and $H_u(w)$ are convex functions of w. Then, the unique roots y_u of (2.11) and (2.12) are measurable functions of u. Hence, the following functions are measurable functions of u:

$$C(u) = C(u, y_u - \alpha)$$

$$C'(u) = C(u, y_u)$$

$$\gamma(u) = \begin{cases} 0 & \text{if } F_\theta^u(C(u) + 0) = F_\theta^u(C(u)) \\ \dfrac{y_u - \alpha - F_\theta^u(C(u))}{F_\theta^u(C(u) + 0) - F_\theta^u(C(u))} & \text{if not} \end{cases}$$

$$\gamma'(u) = \begin{cases} 0 & \text{if } F_{\theta'}^u(C'(u) + 0) = F_{\theta'}^u(C'(u)) \\ \dfrac{y_u - F_{\theta'}^u(C'(u))}{F_{\theta'}^u(C'(u) + 0) - F_{\theta'}^u(C'(u))} & \text{if not.} \end{cases}$$

∎

Theorem 1 *Let Λ be a subset of D'_ν whose interior is nonempty and let Θ_0 and Θ_1 be two disjoint and nonempty subsets of D_ν^1. Let*

$$\Lambda_0 = \{(\theta, \mu) : \theta \in \Theta_0, (\theta, \mu) \in \Lambda\},$$

$$\Lambda_1 = \{(\theta, \mu) : \theta \in \Theta_1, (\theta, \mu) \in \Lambda\}.$$

Then the test Φ^*, which will be defined for three cases, is a strictly u.m.p.u. test of Λ_0 against Λ_1.

Case 1. *When there exists an interior point θ_0 of D_ν^1 such that*

$$\theta_0 \in \bar{\Theta}_0 \cap \bar{\Theta}_1, \qquad \Theta_0 \subset (-\infty, \theta_0[\qquad (\text{or } \Theta_0 \subset]\theta_0, +\infty)),$$

$$\Theta_1 \subset]\theta_0, +\infty) \qquad (\text{or } \Theta_1 \subset (-\infty, \theta_0[),$$

then Φ^* is the right-sided (left-sided) conditional test having level of significance α at the point θ_0.

Case 2. *When $\Theta_1 = \{\theta_1\}$, $\theta_1 \in \bar{\Theta}_0$* (or $\Theta_0 = \{\theta_0\}$, $\theta_0 \in \bar{\Theta}_1$) *and when there exist points of Θ_0* (or (Θ_1) *on each side of θ_1* (or θ_0), *then Φ^** (or $1 - \Phi^*$)

is the two-sided conditional test having level of significance α (or $1-\alpha$) *at the point* θ_1 (or θ_0).

Case 3. *When there exist two points* $\theta_1 < \theta_2$ *belonging to the interior set of* D_v^1, *such that*

$$\theta_1, \theta_2 \in \bar{\Theta}_0 \cap \bar{\Theta}_1;$$

$$[\theta_1, \theta_2] \supset \Theta_1 \text{ (or } \Theta_0);$$

$$\complement]\theta_1, \theta_2[\supset \Theta_0 \text{ (or } \Theta_1).$$

then Φ^* (or $1-\Phi^*$) *is the two-sided conditional test having level of significance* α (or $1-\alpha$) *at the points* θ_1 *and* θ_2.

Proof The proofs for the three cases are similar. Let Φ be an unbiased test. Then we have

Case 1. $\beta_\Phi(\theta_0, \mu) = \alpha$ for every μ, such that $(\theta_0, \mu) \in \Lambda$.

Case 2. $\beta_\Phi(\theta_0, \mu) = \alpha$, $\delta\beta_\Phi(\theta_0, \mu)/\delta\theta = 0$, for every μ, such that $(\theta_0, \mu) \in \Lambda$.

Case 3. $\beta_\Phi(\theta_1, \mu) = \alpha$ for every μ, such that $(\theta_1, \mu) \in \Lambda$ and $\beta_\Phi(\theta_2, \mu) = \alpha$ for every μ, such that $(\theta_2, \mu) \in \Lambda$.

In each of the cases, the set of values of μ has a nonempty interior. Then, for every given θ, we have a complete exponential statistical space with respect to the variable u. Using Theorem 10.2.3 we denote by P_θ^u the conditional distribution given u. Then the previous conditions of unbiasedness are equivalent to:

Case 1. $$\int_{-\infty}^{+\infty} \Phi(t, u)\, dP_{\theta_0}^u(t) = \alpha, \qquad \forall u,$$

Case 2. $$\begin{cases} \displaystyle\int_{-\infty}^{+\infty} t\Phi(t, u)\, dP_{\theta_0}^u(t) = \alpha \int_{-\infty}^{+\infty} t\, dP_{\theta_0}^u(t), & \forall u, \\ \displaystyle\int_{-\infty}^{+\infty} \Phi(t, u)\, dP_{\theta_0}^u(t) = \alpha, & \forall u, \end{cases}$$

Case 3. $$\int_{-\infty}^{+\infty} \Phi(t, u)\, dP_{\theta_1}^u(t) = \int_{-\infty}^{+\infty} \Phi(t, u)\, dP_{\theta_2}^u(t) = \alpha, \qquad \forall u.$$

Now, for every given u, we let $\Phi_u(t) = \Phi(t, u)$ and these conditions become

Case 1. $\beta_{\Phi_u}(\theta_0) = \alpha, \qquad \forall u,$

Case 2. $\beta_{\Phi_u}(\theta_0) = \alpha, \qquad \dfrac{\delta}{\delta\theta}\beta_{\Phi_u}(\theta_0) = 0, \qquad \forall u,$

Case 3. $\beta_{\Phi_u}(\theta_1) = \beta_{\Phi_u}(\theta_2) = \alpha, \qquad \forall u.$

Let us now consider the test $\Phi^*(t, u)$ and let

$$\Phi_u^* = \Phi^*(t, u).$$

By assumption, for every given u, the test Φ_u^* is:

Case 1. the one-sided test having level of significance α at the point θ_0;

Case 2. the two-sided test (or its complement) having level of significance α at the point θ_0;

Case 3. the two-sided test (or its complement) having level of significance α at the points θ_1, θ_2.

Then, by using Property P1 of Lemma 2.5 for Case 1, Lemma 2.4 for Case 2, and Lemma 2.3 for Case 3, we prove that for every given u, the test Φ_u^* is strictly better than Φ_u. By integration with respect to u, we deduce that Φ^* is strictly better than Φ and the proof is completed because Φ^* is an unbiased test. ∎

Exercises

1. Consider the statistical space,

$$\left[\mathbb{N}, \mathscr{B}_{\mathbb{N}}; \mathscr{P}\left(\frac{\theta\mu}{1+\theta}\right), \theta \in \mathbb{R}^+, \mu \in \mathbb{R}^+\right]^n \times \left[\mathbb{N}, \mathscr{B}_{\mathbb{R}}; \mathscr{P}\left(\frac{\mu}{1+\theta}\right), \theta \in \mathbb{R}^+, \mu \in \mathbb{R}^+\right]^n,$$

where $\mathbb{N}$ is the set of nonnegative integers and $\mathscr{P}(\lambda)$ is the Poisson distribution.

(a) Find a sufficient statistic (S, T) such that the conditional distribution of S given $(S + T)$ does not depend on μ.

(b) Find a u.m.p. test of the hypothesis $\{\theta = a\}$ against $\{\theta > a\}$, where a is a given positive number.

2. Consider a sample of size n from the normal distribution $\mathbb{N}(m, \sigma^2)$, where the two parameters are unknown. Find an optimal test of the hypothesis $\{m \leqslant m_0\}$ against $\{m > m_0\}$, where m_0 is a given number. Also find an optimal test of the hypothesis $\{\sigma = \sigma_0\}$ against $\{\sigma \neq \sigma_0\}$, where σ_0 is a given positive number.

3. Consider the statistical space Σ corresponding to the Behrens–Fisher problem (Section 10.5).

(a) Let a be a given number. Find the u.m.p.u. test of the hypothesis $\{\sigma^2/\sigma'^2 \leqslant a\}$ against the hypothesis $\{\sigma^2/\sigma'^2 > a\}$ having level of significance α; also, find a u.m.p.u. test of the hypothesis $\{\sigma^2/\sigma'^2 = a\}$ against $\{\sigma^2/\sigma'^2 \neq a\}$ having level of significance α.

(b) Assume that the ratio σ/σ' is known. Find a u.m.p.u. test of the hypothesis $\{m = m'\}$ against $\{m \neq m'\}$.

(c) Show that in the Behrens–Fisher problem one cannot apply the method of Section 4 and find the only special cases in which this method can be applied.

4. Consider Theorem 3.2. Assume that the condition,

$$\bar{\Theta}_0 \cap \bar{\Theta}_1 = \{\theta_0', \theta_0''\}$$

is not satisfied, and is replaced by

$$\Theta_0 = [\theta_0, \theta_0'], \qquad \Theta_1 = \complement]\theta_1, \theta_1'[, \qquad \text{where} \quad \theta_1 < \theta_0 < \theta_0' < \theta_1'.$$

Show that in this case the u.m.p.u. test of Θ_0 against Θ_1 does not exist.

5. Let $f(u, t)$ be a function on $\mathbb{R}^s \times \mathbb{R}$ such that:

(a) for every t the function $u \to f(u, t)$ is measurable;

(b) for every u the function $t \to f(u, t)$ is a bounded nondecreasing function;

(c) for every u the function $t \to f(u, t)$ is a step function.

Show that $(u, t) \to f(u, t)$ is a measurable function; deduce from it that a function f, satisfying only (a) and (b), is also a measurable function.

6. Show that the test defined in Lemma 4.1 satisfies the condition

$$\lambda = (\theta, \mu), \qquad E_{P_\lambda}(\Phi(T, U) | U) = \alpha.$$

7. Show that the test defined in Lemma 4.2 satisfies the condition:

$$\forall \lambda = (\theta, \mu) \qquad \begin{cases} E_{P_\lambda}(\Phi(T, U) | U) = \alpha, \\ E_{P_\lambda}(T\Phi(T, U) | U) = \alpha E_{P_\lambda}(TU). \end{cases}$$

8. Show that the test defined in Lemma 4.3 satisfies the condition:

$$\forall \lambda = (\theta, \mu) \text{ or } (\theta', \mu), \qquad E_{P_\lambda}(\Phi(T, U) | U) = \alpha.$$

9. Using notation of Section 4, show that

$$G_\theta^u(t) = \int_{-\infty}^{t} z \, dF_\theta^u(z)$$

is a measurable function of (t, u) (Hint: Use the same argument as in Exercise 5). Then, using (1.2), show that $K_u(w)$ is a measurable function of (u, w). Also show that $H_u(w)$ is a measurable function of (u, w).

10. Show how Theorem 5.5.4 can be used in Cases 1 and 3 of Theorem 4.1 to prove that Φ^* is better than any other unbiased test Φ.

11. Using notation of Section 4, suggest a test of the hypothesis $\{\theta \leqslant \theta_0\}$ against $\{\theta \geqslant \theta_1\}$, where $\theta_1 > \theta_0$.

12. Using notation of Section 4, assume that there exists a statistic V which is free for a given value θ_0 of θ. Show that a one-sided test and a two-sided test exist which depend only on V.

13. Consider the statistical space

$$[\mathbb{R}, \mathscr{B}_{\mathbb{R}}; N(m, \sigma_0^2), \quad m \in \mathbb{R}],$$

where $\sigma_0 > 0$ is given. Compute the left-sided test and the two-sided test having level of significance α at the point m_0. Deduce from these the optimal test of the hypothesis $\{m \geqslant m_0\}$ against $\{m \leqslant m_0\}$ and the optimal test of the hypothesis $\{m = m_0\}$ against $\{m \neq m_0\}$.

14. Consider the statistical space

$$[\mathbb{R}^+, \mathscr{B}_{\mathbb{R}^+}; \Gamma(a, \lambda), \qquad \lambda \in \mathbb{R}^+].$$

Compute the left-sided test and the two-sided test having level of significance α at the point λ_0. Also compute the two-sided test having level of significance α at the points λ_0, λ_1.

Apply these results to compute the optimal test of the hypothesis $\{\sigma < \sigma_0\}$ against $\{\sigma > \sigma_0\}$ and the optimal test of the hypothesis $\{\sigma = \sigma_0\}$ against $\{\sigma \neq \sigma_0\}$, defined on the statistical space

$$[\mathbb{R}, \mathscr{B}_{\mathbb{R}}; N(m, \sigma^2), \quad m \in \mathbb{R}, \quad \sigma \in \mathbb{R}^+]^n.$$

15. (Student's tests) Consider the statistical space

$$[\mathbb{R}, \mathscr{B}_{\mathbb{R}}; N(m, \sigma^2), \quad m \in \mathbb{R}, \quad \sigma \in \mathbb{R}^+]^n.$$

Find a u.m.p.u. test of the hypothesis $\{m < m_0\}$ against $\{m > m_0\}$.

Find the u.m.p.u. test of the hypothesis $\{m = m_0\}$ against $\{m \neq m_0\}$ which has level of significance α and compute its power function. In each case apply Theorem 4.1 and, also, give a direct proof using the method of conditional tests given in Section 5.5.

16. Show how Exercise 3(b) can be solved by a suitable conditioning which reduces it to Student's problem.

17. Compute the functions H and K, defined in Lemma 2.5, for a normal distribution and thus, verify the lemma in this special case.

18. Prove Property (P1) of Lemma 2.5.

CHAPTER 12

Functional Analysis and Mathematical Statistics

We have seen that certain problems of mathematical statistics are related to various problems of functional analysis. Some examples of this relationship will now be given. A better understanding of the results may lead to further productive research.

1. Computation of a Statistic Having a Given Image

The image of a statistic defined on an exponential statistical space can be computed by means of the Laplace transform (see Exercise 10.13). We now give some results for the converse problem.

Definition 1 Let f be a real-valued function defined on $(\mathbb{R}^s, \mathscr{B}_{\mathbb{R}^s})$. If the measure m defined by $dm = f\,dx$ is a smooth measure on $(\mathbb{R}^s, \mathscr{B}_{\mathbb{R}^s})$ (Definition 10.1.3), we say that f is a *smooth function*. We call $l_f(z)$ the *Laplace transform* of f, where

$$l_f(z) = L_m(z), \qquad \text{Re}(z) \in D_f = D_m.$$

The Laplace transform $L_m(z)$ of the measure m was defined in Section

10.1. When $s = 1$, we obtain some known results. If f is defined on $\mathbb{R}^+$, then l_f is the following one-sided Laplace transform

$$l_f(z) = \int_0^\infty e^{-zx} f(x)\, dx,$$

where this integral is absolutely convergent when $\mathrm{Re}(z)$ is greater than some θ_0. If f is defined on $\mathbb{R}$, then l_f is the following two-sided Laplace transform

$$l_f(z) = \int_{-\infty}^{+\infty} e^{-zx} f(x)\, dx,$$

and this integral is absolutely convergent when $\mathrm{Re}(z)$ belongs to some interval D_f, which may be empty.

Theorem 1 *Let m_1 and m_2 be two smooth measures on $(\mathbb{R}^s, \mathscr{B}_{\mathbb{R}^s})$ such that D_{m_1} and D_{m_2} have a nonempty intersection. If m_1 is absolutely continuous with respect to Lebesgue measure*

$$dm_1 = f\, dx,$$

then $m_1 * m_2$ is also absolutely continuous with respect to Lebesque measure and we have

$$\frac{d(m_1 * m_2)}{dx} = \int_{\mathbb{R}^s} f(x - \xi)\, dm_2(\xi) \qquad \text{a.e.}$$

Moreover, if m_2 is also absolutely continuous with respect to Lebesgue measure

$$dm_2 = g\, dx,$$

then we have, almost everywhere,

$$\frac{d(m_1 * m_2)}{dx} = f * g = \int_{\mathbb{R}^s} f(x - \xi) g(\xi)\, d\xi = \int_{\mathbb{R}^s} f(\xi) g(x - \xi)\, d\xi.$$

Proof Let us consider the measure μ defined by

$$d\mu = \left[\int_{\mathbb{R}^s} f(x - \xi)\, dm_2(\xi) \right] dx.$$

We have

$$L_\mu(z) = \int_{\mathbb{R}^s} e^{-\langle z | x \rangle} \left[\int_{\mathbb{R}^s} f(x - \xi)\, dm_2(\xi) \right] dx.$$

Furthermore, if $\mathrm{Re}(z) \in D_{m_1} \cap D_{m_2}$ then the following two integrals

$$L_{m_2}(z) = \int_{\mathbb{R}^s} e^{-\langle z|\xi\rangle} \, dm_2(\xi),$$

$$l_f(z) = \int_{\mathbb{R}^s} e^{-\langle z|\eta\rangle} f(\eta) \, d\eta$$

exist. Thus the function $e^{-\langle z|\xi+\eta\rangle} f(\eta)$ is integrable with respect to the measure $dm_2(\xi)\, d\eta$ on $\mathbb{R}^s \times \mathbb{R}^s$. Now, we consider the transformation

$$(\xi, \eta) \leftrightarrows (x = \xi + \eta, \xi)$$

and apply Fubini's Theorem (Theorem A.6.2). Then the function

$$k(x) = \int_{\mathbb{R}^s} f(x - \xi) \, dm_2(\xi)$$

is defined almost everywhere and we have

$$L_\mu(z) = L_{m_2}(z) l_f(z) = L_{m_1}(z) L_{m_2}(z).$$

The Laplace transforms of the measures μ and $m_1 * m_2$ are the same if

$$\mathrm{Re}(z) \in D_{m_1} \cap D_{m_2}.$$

Then, using the remark after Theorem 10.1.3, we see that these measures are equal.

The second part of the theorem follows easily from the first part. The *function* $f * g$ is called the *convolution* of f and g and it is defined if D_{m_1} and D_{m_2} have a nonempty intersection. Moreover, for every z such that

$$\mathrm{Re}(z) \in D_{m_1} \cap D_{m_2}$$

we have

$$l_f(z) l_g(z) = l_{f*g}(z).$$

Theorem 2 *Let f be a smooth function and let θ be a point of D_f. If the function $t \to |l_f(\theta - it)|$ is integrable on $\mathbb{R}^s$, then*

$$f(x) = (2\pi)^{-s} \int_{\mathbb{R}^s} l_f(\theta - it) e^{\langle x|\theta - it\rangle} \, dt \qquad \text{a.e.} \tag{1}$$

Proof We can restrict ourselves to the case where f is positive. Then the function

$$g(x) = \frac{f(x) e^{-\langle x|\theta\rangle}}{l_f(\theta)}, \qquad x \in \mathbb{R}^s$$

is a probability density function whose characteristic function is

$$\varphi(t) = \frac{l_f(\theta - it)}{l_f(\theta)}, \qquad t \in \mathbb{R}^s.$$

By assumption $\varphi(t)$ is integrable on $\mathbb{R}^s$. Then using Levy's Theorem on the inversion of the characteristic function (see, e.g., Luckas and Laha [46, p. 24]), we have

$$g(x) = (2\pi)^{-s} \int_{\mathbb{R}^s} e^{-i\langle t|x\rangle} \varphi(t)\, dt. \qquad \blacksquare$$

Remark The following formula, which is analogous to (1), also holds for iterated integrals of complex variables

$$f(x) = (2\pi i)^{-s} \int_{\theta_1 - i\infty}^{\theta_1 + i\infty} \cdots \int_{\theta_s - i\infty}^{\theta_s + i\infty} l_f(z) e^{\langle x|z\rangle}\, dz.$$

We note that, since the function $l_f(z)$ is analytic when $\mathrm{Re}(z) \in D_f^0$, we can replace the path of integration by any other regular path.

Theorem 3 *Let Δ be the subset of $\mathbb{R}^s$ defined by*

$$\theta = (\theta_1, \ldots, \theta_s), \qquad A_j \leqslant \theta_j < B_j, \qquad j = 1, \ldots, s,$$

where the constants A_j are bounded, but the constants B_j can be infinite. Let φ be an analytic function on an open set which contains the set

$$B = \{\mathrm{Re}(z) \in \Delta\}.$$

If there exists a constant K such that

$$|\varphi(\theta + it)| \leqslant \frac{K}{(1 + t_1^2)\cdots(1 + t_s^2)}, \qquad \forall \theta \in \Delta, \quad \forall t = (t_1, \ldots, t_s) \in \mathbb{R}^s, \quad (2)$$

then φ is the Laplace transform of a function f such that

(i) *the interior set of Δ is included in D_f,*

(ii) $|f(x)| \leqslant K' e^{\langle x|\theta\rangle}, \qquad \forall \theta \in \Delta.$

Furthermore, if $B_j = +\infty$, we have

$$\forall x = (x_1, \ldots, x_s), \qquad x_j < 0 \;\Rightarrow\; f(x) = 0.$$

Proof We can restrict ourselves to the case where the constants A_j are zero. Condition (2) implies that the integral

$$\mathscr{I}_\theta(x) = \int_{\mathbb{R}^s} \varphi(\theta + it) e^{i\langle x|t\rangle}\, dt$$

is uniformly convergent and its absolute value is uniformly bounded. Now by induction on s we prove that the function

$$f_\theta(x) = \frac{e^{\langle x|\theta\rangle}}{(2\pi)^s}\mathscr{I}_\theta(x) = (2\pi)^{-s}\int_{\mathbb{R}^s}\varphi(\theta+it)e^{\langle x|\theta+it\rangle}\,dt \tag{3}$$

does not depend on θ. When $s = 1$, this property is deduced from Cauchy's theorem applied to the contour in Fig. 1. The function $\varphi(z)e^{xz}$ is analytic in the strip

$$0 < \operatorname{Re}(z) < B_1.$$

Then

$$f_{\theta_1}(x) - f_{\theta_2}(x) = \lim_{R\to+\infty}\int_{\theta_1}^{\theta_2}[\varphi(\theta - iR)e^{(\theta-iR)x} + \varphi(-\theta+iR)e^{(-\theta+iR)x}]\,d\theta$$

and

$$|f_{\theta_1}(x) - f_{\theta_2}(x)| \leqslant \lim_{R\to+\infty}(2\pi)^{-s}\frac{K}{R^2}\int_{\theta_1}^{\theta_2}e^{\theta x}\,d\theta = 0.$$

Now we assume that the function $\psi(x_1,\ldots,x_{s-1};\theta_s+it_s)$, which is equal to

$$(2\pi)^{-s+1}\int_{\mathbb{R}^{s-1}}\varphi(\theta+it)\exp\left\{\sum_{j=1}^{s-1}x_j(\theta_j+it_j)\right\}dt_1\cdots dt_{s-1},$$

does not depend on $\theta_1,\ldots,\theta_{s-1}$. This function is analytic, as a function of the variable $\theta_s + it_s$, and satisfies the assumptions of the theorem when $s = 1$. Then the integral,

$$\frac{1}{2\pi}\int_{-\infty}^{+\infty}\psi(x_1,\ldots,x_{s-1};\theta_s+it_s)e^{x_s\theta_s+ix_st_s}\,dt_s$$

which is equal to $f_\theta(x)$, does not depend on θ. We designate this function by $f(x)$ and define it by (3) with $\theta = 0$.

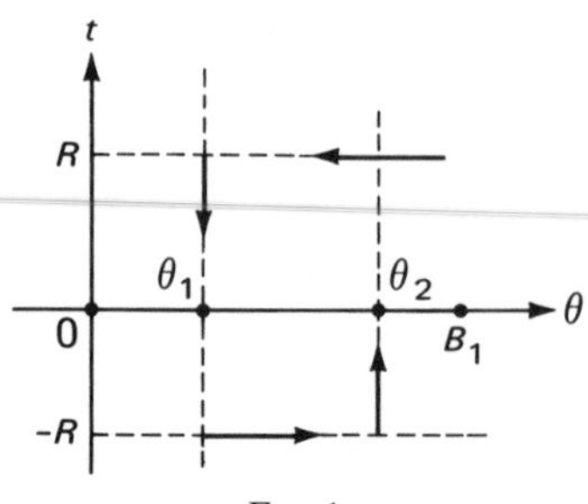

FIG. 1

Using (3) and the boundness of $\mathscr{J}_\theta$, we have

$$|f(x)| \leqslant K' e^{\langle x|\theta\rangle}, \qquad \forall\theta \in \Delta. \tag{4}$$

Let θ^0 be an interior point of Δ. There exists a real $\varepsilon > 0$ such that

$$0 \leqslant \theta_j = \theta_j^0 - \varepsilon\, \mathrm{sign}(x_j) < B_j, \qquad \forall j = 1, \ldots, s.$$

Using (4) with $\theta = (\theta_1, \ldots, \theta_s)$, we can prove that θ^0 belongs to D_f. Then D_f contains the interior set of Δ.

The relation

$$\mathscr{J}_{\theta^0}(x) = (2\pi)^s f(x) e^{-\langle x|\theta^0\rangle} \tag{5}$$

shows that $\mathscr{J}_{\theta^0}(x)$ is integrable. By definition it can be considered as the Fourier transform of the complex function $t \to \varphi(\theta^0 + it)$. Then we can apply Theorem 2 to show that

$$\varphi(\theta^0 + it) = (2\pi)^{-s} \int_{\mathbb{R}^s} \mathscr{J}_{\theta^0}(x) e^{-i\langle t|x\rangle}\, dx.$$

Now, using (5) we have

$$\varphi(\theta^0 + it) = \int_{\mathbb{R}^s} f(x) e^{-\langle \theta^0 + it|x\rangle}\, dx.$$

This implies that for every $z \in B$, $\varphi(z)$ is the Laplace transform of f.

Finally, if for some integer j $(1 \leqslant j \leqslant s)$, B_j equals $+\infty$, we consider (4) with $x_j < 0$. Then θ_j can increase to infinity, whence $|f(x)|$ vanishes. ∎

2. Sufficient Statistics Having Minimum Dimension

This section continues, by means of functional analysis, the construction of minimum sufficient statistics, begun in Section 2.5. First, we recall some basic results which are needed in the last part of this section. A mapping f, from a metric space E onto a metric space F, is a *homeomorphism* [13, p. 47] if f is a one-to-one mapping and if f and f^{-1} are continuous. In particular $\mathbb{R}^n$ and $\mathbb{R}^m$ are *homeomorphic spaces* if and only if $n = m$ [16, p. 359].

Theorem 1 *Let A and A' be two open sets of $\mathbb{R}^n$, let $\mathscr{B}$ be the σ-field of Borel subsets of A and let f be a homeomorphism from A' onto A. Then $f^{-1}(\mathscr{B})$ is the σ-field of the Borel sets of A'. Furthermore, if f is continuously differentiable, then we have*

$$\mathscr{B}_1, \mathscr{B}_2 \subset \mathscr{B}, \qquad f^{-1}(\mathscr{B}_1) \overset{L}{=} f^{-1}(\mathscr{B}_2) \quad \Rightarrow \quad \mathscr{B}_1 \overset{L}{=} \mathscr{B}_2,$$

where L is Lebesgue measure on $\mathbb{R}^n$.

Proof The function f is continuous. Then the σ-field $\mathscr{B}'$ of the Borel sets of A' contains the subfield $f^{-1}(\mathscr{B})$. On the other hand, if B is an open set of A', then $f(B)$ is an open set of A and we have

$$f^{-1}(f(B)) = B.$$

Then every open subset of A' belongs to the σ-field $f^{-1}(\mathscr{B})$ which is finally equal to the subfield $\mathscr{B}'$.

Now, if $f^{-1}(\mathscr{B}_1)$ and $f^{-1}(\mathscr{B}_2)$ are L-equal, using Theorem 2.5.1 we have

$$\forall B_1 \in \mathscr{B}_1, \exists B_2 \in \mathscr{B}_2 \qquad \text{such that} \qquad L(f^{-1}(B_1) \triangle f^{-1}(B_2)) = 0.$$

But f is an onto mapping and we have

$$f(f^{-1}(B_1) \triangle f^{-1}(B_2)) = B_1 \triangle B_2.$$

On the other hand, since f is continuously differentiable, we have [54, p. 15]

$$L(B_1 \triangle B_2) = 0. \qquad \blacksquare$$

Theorem 2 [13, p. 273] *Let f be a continuously differentiable mapping from a neighborhood Ω of a ($a \in \mathbb{R}^n$) into $\mathbb{R}^n$. If the rank of the derivative of f at the point a is n, there exist a neighborhood U of a and a neighborhood V of $f(a)$ such that the mapping f, from U onto V, is a homeomorphism.*

Theorem 3 [13, p. 277] *Let f be a continuously differentiable mapping from a neighborhood Ω of a ($a \in \mathbb{R}^n$) into $\mathbb{R}^m$ such that the rank of the derivative of f is equal to r at every point of Ω. Then there exist*

(1) *an open neighborhood U of a ($U \subset \Omega$) and a continuously differentiable homeomorphism from U onto the unit ball I^n ($|x_i| < 1$; $i = 1, \ldots, n$) of $\mathbb{R}^n$.*

(2) *an open neighborhood V of $f(a)$ ($V \supset f(U)$ and a continuously differentiable homeomorphism v from the unit ball I^m ($|y_j| < 1$; $j = 1, \ldots, m$) of $\mathbb{R}^m$ onto V, such that*

$$f = v(f_0(u)),$$

where f_0 is the following mapping of I^n into I^m,

$$(x_1, \ldots, x_n) \rightarrow (x_1, \ldots, x_r, 0, \ldots, 0).$$

The previous results are used to study sufficient statistics having minimum dimension (see also [15], [9], [41, p. 35]).

Definition 1 Let $(\mathbb{R}^n, \mathscr{B}_{\mathbb{R}^n}, \mathscr{P})$ be a dominated statistical space. We say that it is *irreducible* if there exists a $\mathscr{P}$-minimum subfield $\mathscr{B}$ and an open subset Ω_0 of $\mathbb{R}^n$ such that the following two conditions, involving a Borel set B of Ω_0, are equivalent:

(i) the Lebesgue measure of B is zero;
(ii) B belongs to $\Omega_0 \cap \overline{\mathscr{B}}$ and it is $\mathscr{P}$-negligible.

In this definition, following Theorem 2.5.1 we have designated by $\overline{\mathscr{B}}$ the smallest subfield which contains $\mathscr{B}$ and the $\mathscr{P}$-negligible events. On the other hand we have designated by $\Omega_0 \cap \overline{\mathscr{B}}$ the subfield of the sets $\{\Omega_0 \cap B, B \in \overline{\mathscr{B}}\}$.

Theorem 4 *Let $(\mathbb{R}^n, \mathscr{B}_{\mathbb{R}^n}, \mathscr{P})$ be an irreducible statistical space, then every continuously differentiable sufficient statistic has at least n components.*

Proof Let X be a continuously differentiable sufficient statistic taking values in $(\mathbb{R}^m, \mathscr{B}_{\mathbb{R}^m})$ and let r be the maximum on Ω_0 of the rank of the derivative X' of X. Every minor of X' is a continuous function. Thus the maximum of the rank of X' is reached on an open subset Ω_0 of Ω. Now, we can apply Theorem 3. We designate by $\tilde{X}$ the restriction on U of the mapping X.

The subfield $\mathscr{B}$ is $\mathscr{P}$-minimum and sufficient. Thus we have

$$\mathscr{B}_{\mathbb{R}^n} \supset \overline{X^{-1}(\mathscr{B}_{\mathbb{R}^m})} \supset \overline{\mathscr{B}}.$$

Denoting by $\mathscr{B}_U$ the σ-field of the Borel sets of U, we have

$$\mathscr{B}_U \supset U \cap \overline{X^{-1}(\mathscr{B}_{\mathbb{R}^m})} \supset U \cap \overline{\mathscr{B}} = \mathscr{B}_U.$$

Then

$$\mathscr{B}_U = U \cap \overline{X^{-1}(\mathscr{B}_{\mathbb{R}^m})} \stackrel{\mathscr{P}}{=} U \cap X^{-1}(\mathscr{B}_{\mathbb{R}^m}).$$

But on Ω_0, the notion of $\mathscr{P}$-equivalence is the same as that of L-equivalence and we have

$$\mathscr{B}_U \stackrel{L}{=} U \cap X^{-1}(\mathscr{B}_{\mathbb{R}^m}) = \tilde{X}^{-1}(\mathscr{B}_V),$$

where $\mathscr{B}_V$ is the subfield of the Borel sets of V. The relation

$$\tilde{X} = v(f_0(u)),$$

and Theorem 1 show that, using obvious notation, we have

$$v^{-1}(\mathscr{B}_V) = \mathscr{B}_{I^m},$$

$$\mathscr{B}_{I^n} \stackrel{L}{=} f_0^{-1}(\mathscr{B}_{I^m}) = \mathscr{B}_{I^r} \times I^{n-r}.$$

Then

$$n = r \leqslant m. \qquad \blacksquare$$

Theorem 5 *Let $f_1, \ldots, f_k$ be k real-valued and continuously differentiable functions defined on an interval Ω of $\mathbb{R}$. If the functions 1, $f_1, \ldots, f_k$ are linearly independent, then for every integer n greater than or equal to k the derivative of the mapping g from Ω^n in $\mathbb{R}^k$,*

$$x_i \in \Omega, \qquad g_i(x_1, \ldots, x_n) = \sum_{j=1}^{n} f_i(x_j), \qquad i = 1, \ldots, k,$$

has the rank k on a nonempty open subset of Ω^n.

The proof of this theorem is left as Exercise 6.

Theorem 6 *Consider the exponential statistical space,*

$$\left[\mathbb{R}^s, \mathscr{B}_{\mathbb{R}^s}; \frac{dP_\theta}{dx} = \frac{e^{\langle\theta|x\rangle}}{l_f(-\theta)} f(x), \quad \theta \in \Theta \subset D'_f\right].$$

If the dimension of Θ is equal to s and if f is continuous, then this statistical space is irreducible.

Proof Using Theorem 10.3.2, we prove that $\mathscr{B}_{\mathbb{R}^s}$ is a $\mathscr{P}$-minimum sufficient subfield. On the other hand, since f is continuous there exists an open subset on which f is strictly positive. Then the conditions of Definition 1 are satisfied (see Exercise 3). ∎

Theorem 7 *Let I be an interval of $\mathbb{R}$, let $\mathscr{B}_I$ be the σ-field of the Borel sets of I and let $(I, \mathscr{B}_I;\ p_\theta,\ \theta \in \Theta)$ be a statistical space dominated by Lebesgue measure on I. If the following conditions are satisfied:*

(i) *for every θ, the function $x \to p_\theta(x)$ is continuously differentiable,*
(ii) *there exists $\theta_0 \in \Theta$ such that p_{θ_0} is strictly positive on I,*
(iii) *the integer n is less than the dimension d of the linear space $\mathscr{L}$ generated by 1 and the functions $\log(p_\theta(\omega)/p_{\theta_0}(\omega))$ where θ runs through Θ,*

then the following statistical space is irreducible:

$$(I, \mathscr{B}_I; p_\theta, \quad \theta \in \Theta)^n.$$

Proof Let $\{1, f_j, j \in \mathscr{J}\}$ be a basis of $\mathscr{L}$ (d may be infinite). Using Theorem 2.5.6, we see that the statistic X,

$$x_i \in I, \qquad X : (x_1, \ldots, x_n) \to \left\{\sum_{i=1}^{n} f_j(x_i), \quad j \in \mathscr{J}\right\}$$

is $\mathscr{P}$-minimum, sufficient. Since $n < d$, without loss of generality, we can assume that

$$\mathscr{J} \supset [1, 2, \ldots, n].$$

Then the subfield which is induced by X on I^n contains the subfield which is induced by the functions,

$$\left\{\sum_{i=1}^{n} f_j(x_i), \quad j = 1, \ldots, n\right\}.$$

Now we can apply Theorem 5 and Theorem 2 and verify (Exercises 4 and 5) that the conditions of Definition 1 are satisfied. ∎

3. Spaces of Statistics

First we recall some results involving the spaces L_p (see, e.g., Neveu [53, p. 55]). Let $(\Omega, \mathscr{A}, P)$ be a probability space. The two random variables X and X' are said to be equivalent if

$$P(X \neq X') = 0.$$

With every random variable we associate the real numbers,

$$\|X\|_p = [E(|X|^p)]^{1/p}, \qquad 1 \leqslant p < \infty,$$

$$\|X\|_\infty = \sup[x \mid P(|X| > x) > 0].$$

These numbers may be infinite and depend only on the equivalence class of X.

Theorem 1 *For every real number p ($p \geqslant 1$) we designate by $L_p(\Omega, \mathscr{A}, P)$ the linear space of the equivalence class of random variables on $(\Omega, \mathscr{A}, P)$, such that*

$$\|X\|_p < \infty.$$

This space L_p is a Banach space for this norm.

For every real p and q such that

$$1 \leqslant p < \infty \qquad \text{and} \qquad \frac{1}{p} + \frac{1}{q} = 1;$$

L_q is the dual space of L_p [53, p. 113]. Moreover, L_∞ is the dual space of L_1; but in general, L_1 is not the dual space of L_∞. We define $\sigma(L_\infty, L_1)$, *the weak topology induced by L_1 on L_∞*, as the weakest topology for which all linear forms F_X on L_∞, defined by

$$\forall X \in L_1, \qquad F_X : Y \to E(XY)$$

are continuous. A fundamental system of neighborhoods of the origin is given by the subsets

$$V_{\varepsilon, n, X_1, \ldots, X_n} = \{Y \in L_\infty : |E(X_i Y)| \leqslant \varepsilon, \quad i = 1, \ldots, n\}$$

for every $\varepsilon > 0$, every integer n and every system $\{X_1, \ldots, X_n\}$ of functions belonging to L_1.

Definition 1 Let $(\Omega, \mathscr{A})$ be a probability space. We denote by

$$\mathscr{L}_\infty(\Omega, \mathscr{A}) \qquad (\text{or } (\mathscr{M}(\Omega, \mathscr{A}))),$$

the linear space of the bounded measurable real-valued functions (or bounded real measures) on $(\Omega, \mathscr{A})$.

Every bounded measure m is the difference between two positive bounded measures m^+ and m^-. If we introduce on $\mathscr{M}(\Omega, \mathscr{A})$, the norm

$$\|m\| = m^+(\Omega) + m^-(\Omega)$$

then $\mathscr{M}$ is a Banach space [53, p. 107]. We define $\sigma\{\mathscr{L}_\infty, \mathscr{M}\}$, *the weak topology induced by* $\mathscr{M}$ *on* $\mathscr{L}_\infty$ as the weakest topology for which all linear forms on $\mathscr{L}_\infty$, defined by

$$\forall \mu \in \mathscr{M}, \qquad F_\mu : f \to \int_\Omega f \, d\mu = \int_\Omega f \, d\mu^+ - \int_\Omega f \, d\mu^-,$$

are continuous. In this weak topology, a fundamental system of neighborhoods of the origin is given by the subsets

$$W_{\varepsilon, n, \mu_1, \ldots, \mu_n} = \left\{ f \in \mathscr{L}_\infty : \left| \int_\Omega f \, d\mu_i \right| \leqslant \varepsilon, \quad i = 1, \ldots, n \right\}$$

for every $\varepsilon > 0$, for every integer n and every system $\{\mu_1, \ldots, \mu_n\}$ of measures belonging to $\mathscr{M}$.

Now, consider the statistical space $(\Omega, \mathscr{A}, \mathscr{P})$. We shall also add the index $P \in \mathscr{P}$ in writing the norms since the probability distribution runs through $\mathscr{P}$.

Definition 2 For every real number p $(p > 1)$ we call $L_p(\Omega, \mathscr{A}, \mathscr{P})$ (or $\Lambda_p(\Omega, \mathscr{A}, \mathscr{P})$) the space of the $\mathscr{P}$-equivalence classes of statistics on $(\Omega, \mathscr{A}, \mathscr{P})$ such that

$$\forall P \in \mathscr{P}, \qquad \|X\|_p^P < \infty, \qquad \left[\text{or } \sup_{P \in \mathscr{P}}(\|X\|_p^P) < \infty\right].$$

We remark that for every $P \in \mathscr{P}$, $\mathscr{P}$-equivalence implies P-equivalence. Thus $\|X\|_p^P$ is well defined on a $\mathscr{P}$-equivalence class X. One can prove that the spaces Λ_p are Banach spaces for the norms

$$\sup_{P \in \mathscr{P}} \| X \|_p^P.$$

Furthermore, using the remark following Definition 1.2.4, one can prove easily the following

Theorem 2 *If the statistical space* $(\Omega, \mathscr{A}, \mathscr{P})$ *is dominated by a special distribution* P^* *then we have*

$$\Lambda_\infty(\Omega, \mathscr{A}, \mathscr{P}) = L_\infty(\Omega, \mathscr{A}, \mathscr{P}^*) \subset L_\infty(\Omega, \mathscr{A}, \mathscr{P}).$$

Theorem 3 *If the statistical space* $(\Omega, \mathscr{A}, \mathscr{P})$ *is dominated by a special distribution* P^*, *then every family* $\mathscr{K}$ *of statistics, satisfying*

$$\exists X_0 \in \Lambda_\infty, \qquad b \in \mathbb{R}^+, \qquad \text{such that} \qquad \forall X \in \mathscr{K}, \qquad \| X - X_0 \|_\infty^{P^*} \leqslant b,$$

is a compact subset of $\Lambda_\infty(\Omega, \mathscr{A}, \mathscr{P})$ *for the topology* $\sigma(L_\infty(\Omega, \mathscr{A}, P^*), L_1(\Omega, \mathscr{A}, P^*))$.

This theorem is deduced from Theorem 2 and the Theorem of Banach–Alaoglu (see, e.g., Wilansky [66, p. 239]).

Theorem 4 *Let* $[\Omega, \mathscr{A}, \mathscr{P} = \{P_\theta, \theta \in (\Theta, \mathscr{T})\}]$ *be a statistical space which is dominated by the special distribution* P^*. *If we introduce on* $L_\infty(\Omega, \mathscr{A}, P^*)$ *the topology* $\sigma(L_\infty(\Omega, \mathscr{A}, P^*), L_1(\Omega, \mathscr{A}, P^*))$ *and on* $\mathscr{L}_\infty(\Theta, \mathscr{T})$ *the topology* $\sigma\{\mathscr{L}_\infty(\Theta, \mathscr{T}), \mathscr{M}(\Theta, \mathscr{T})\}$, *then the mapping* β, *defined in Section* 1.4, *is a continuous mapping from* Λ_∞ *to* $\mathscr{L}_\infty$.

Proof It is sufficient to prove the continuity at the origin. We use the notations of Section 1.4 and the definitions of weak topologies. Consider $m \in \mathscr{M}(\Theta, \mathscr{T})$ and $X \in \Lambda_\infty(\Omega, \mathscr{A}, \mathscr{P})$. The function β_X is bounded. Then it is integrable with respect to the positive and the negative parts of m. Now, using Theorems 1.3.2 and A.6.1, we have

$$\begin{aligned} \int_\Theta \beta_X \, dm &= \int_\Theta \beta_X \, dm^+ - \int_\Theta \beta_X \, dm^- \\ &= \int_\Omega X\left(\frac{d\beta^*_{m^+}}{dP^*}\right) dP^* - \int_\Omega X\left(\frac{d\beta^*_{m^-}}{dP^*}\right) dP^* = \int_\Omega X \frac{d\beta^*_m}{dP^*} \, dP^*, \end{aligned}$$

where

$$\frac{d\beta^*_m}{dP^*} \in L_1(\Omega, \mathscr{A}, P^*).$$

Then

$$\left\{\left|\int_\Theta \beta_X\, dm_i\right| \leqslant \varepsilon,\quad i = 1, \ldots, n\right\} \Leftrightarrow$$

$$\left\{\left|\int_\Omega X\left(\frac{d\beta^*_{m_i}}{dP^*}\right) dP^*\right| \leqslant \varepsilon,\quad i = 1, \ldots, n\right\}. \quad \blacksquare$$

From Theorems 3 and 4 we can deduce:

Theorem 5 *The assumptions of Theorem 3 imply that the image of* $\mathscr{K}$ *by the mapping* β *is a subset of* $\mathscr{L}_\infty(\Theta, \mathscr{T})$, *which is compact with respect to the topology* $\sigma\{\mathscr{L}_\infty, \mathscr{M}\}$.

4. Existence of Free Events

In some statistical problems, in particular hypothesis-testing and estimation, one seeks set indicators which belong to some linear subspace V of $L_\infty(\Omega, \mathscr{A}, \mathscr{P})$. For example, we shall consider the linear subspaces,

$$V_1 = \{X : E_P(X) = E_P(X_0),\quad \forall P \in \mathscr{P}\}$$

$$V_2 = \{X : E_P(X \mid \mathscr{B}) = E_P(X_0 \mid \mathscr{B}),\quad \forall P \in \mathscr{P}\},$$

where X_0 is a test and $\mathscr{B}$ a subfield of $\mathscr{A}$. We shall also consider the linear subspace (Section 2.7)

$$V_3 = \{X : E_P(X \mid \mathscr{B}_j) = E_P(X_0 \mid \mathscr{B}_j),\quad \forall P \in \mathscr{P},\quad j \in J\}$$

where $(\mathscr{B}_j, j \in J)$ is a finite family of subfields of $\mathscr{A}$.

Definition 1 Let $(\Omega, \mathscr{A}, \mathscr{P})$ be a statistical space dominated by a special distribution P^*. We say that the subset W of $L_1(\Omega, \mathscr{A}, P^*)$ is *thin* if for every $A \in \mathscr{A}$ such that $P^*(A) > 0$, there exists a bounded real-valued statistic X such that

$$P^*(X \neq 0) > 0, \qquad P^*(X1_{\complement A} \neq 0) = 0, \qquad \int_\Omega ZX\, dP^* = 0, \qquad \forall Z \in W.$$

We denote by W' the subset of $L_\infty(\Omega, \mathscr{A}, P^*)$,

$$W' = \left\{X : \int_\Omega XZ\, dP^* = 0,\quad \forall Z \in W\right\}.$$

Theorem 1 *Let* $\mathscr{T}$ *be the ball of* $L_\infty(\Omega, \mathscr{A}, P^*)$ *which is defined by*

$$0 \leqslant X \leqslant 1, \qquad P^*\text{-a.e.}$$

The subset W is thin if and only if, for every $X_0 \in \mathscr{T}$, all the extremal points of the convex set

$$K = \mathscr{T} \cap (X_0 + W')$$

are equivalence classes of set indicators.

Proof Using Theorem 3.3 we can prove that the set $\mathscr{T}$ is convex and compact with respect to the topology $\sigma(L_\infty, L_1)$. Then K is also convex and compact and since it is nonempty, it has extremal points (Krein–Millman Theorem). (See Bourbaki [10].)

Now we shall prove the sufficient condition, by *reductio ad absurdum*. Let X^* be an extremal point of K which does not contain a set indicator, then there exists a positive number ε such that

$$P^*(X^*(1 - X^*) > \varepsilon) > 0.$$

But W is a thin set. Thus there exists a statistic $\tilde{X}$, not identically zero, which belongs to W' and vanishes on the set

$$\{X^*(1 - X^*) \leqslant \varepsilon\}.$$

Then the two statistics

$$X^* + \varepsilon \frac{\tilde{X}}{\|\tilde{X}\|_\infty}, \qquad X^* - \varepsilon \frac{\tilde{X}}{\|\tilde{X}\|_\infty},$$

belong to K and are different from each other and from X^*. This is impossible.

Conversely, if W is not thin, there exists an event $\tilde{A} \in \mathscr{A}$ which is P^*-negligible, such that

$$X \in W' - \{0\} \quad \Rightarrow \quad P^*(X 1_{\complement \tilde{A}} \neq 0) > 0. \tag{1}$$

The proof is completed if we show, again by *reductio ad absurdum*, that the equivalence class Z of $\frac{1}{2} 1_{\tilde{A}}$ is an extremal point of the convex set,

$$K_0 = \mathscr{T} \cap (Z + W').$$

However, two distinct elements X_1 and X_2 exist in K_0, such that

$$Z = \frac{X_1 + X_2}{2}.$$

Thus we have

$$P^*(X_1 1_{\complement \tilde{A}} \neq 0) = 0, \qquad P^*(X_2 1_{\complement \tilde{A}} \neq 0) = 0,$$
$$X = Z - X_1 = X_2 - Z \in W'.$$

But this is inconsistent with (1). ∎

Definition 2 Let μ be a positive measure on the measurable space $(\Omega, \mathscr{A})$. Every equivalence class of $L_\infty(\Omega, \mathscr{A}, \mu)$ which contains the indicator of a set A such that

$$\mu(A) > 0, \qquad \forall B \in \mathscr{A} : [B \subset A \quad \Rightarrow \quad \mu(B)\mu(A - B) = 0]$$

is called an *atom of* $(\Omega, \mathscr{A}, \mu)$.

For instance, a probability distribution on $(\mathbb{R}^n, \mathscr{B}_{\mathbb{R}^n})$ which is absolutely continuous with respect to Lebesgue measure does not have atoms (is nonatomic).

Theorem 2 (Liapunov) *Let $P_1, \ldots, P_N$ be N nonatomic probability distributions on $(\Omega, \mathscr{A})$. Then, for every test X, there exist a set $A \in \mathscr{A}$ such that*

$$P_i(A) = E_{P_i}(X), \qquad \forall i = 1, \ldots, N.$$

Proof Let P^* be the distribution,

$$P^* = \frac{P_1 + \cdots + P_N}{N}.$$

The theorem is proved if we show that the subset W of $L_1(\Omega, \mathscr{A}, P^*)$, which is the union of the N equivalence classes

$$dP_j/dP^*, \qquad j = 1, \ldots, N,$$

is thin. More generally, we shall prove by induction on N, the number of elements belonging to W, that if P^* is nonatomic, then every finite set W is thin. Let $Z_1, \ldots, Z_N$ be the elements of W and let A be an event of $\mathscr{A}$ such that $P^*(A) > 0$. Since P^* is nonatomic, there exists a subset A_1 of A, such that

$$A_1 \in \mathscr{A}, \quad A - A_1 = A_2 \in \mathscr{A}, \qquad P^*(A_1) > 0, \quad P^*(A_2) > 0.$$

If $\{Z_1, \ldots, Z_{N-1}\}$ is a thin set, there exist two statistics X_1 and X_2, such that

$$P^*(X_1 \neq 0) > 0, \qquad P^*(X_2 \neq 0) > 0,$$

$$P^*(X_1 1_{\complement A_1} \neq 0) = 0, \qquad P^*(X_2 1_{\complement A_2} \neq 0) = 0,$$

$$\int_\Omega Z_j X_1 \, dP^* = \int_\Omega Z_j X_2 \, dP^* = 0, \qquad j = 1, \ldots, N - 1.$$

But there exist two nonzero constants α and β such that

$$\int_\Omega (\alpha X_1 + \beta X_2) Z_N \, dP^* = 0.$$

Then $\{Z_1, \ldots, Z_N\}$ is a thin set. We can show by an analogous argument that $\{Z_1\}$ is thin. ∎

Theorem 3 (Romanovski–Sudakov) *Let $P_1, \ldots, P_N$ be N distributions on $(\mathbb{R}^2, \mathscr{B}_{\mathbb{R}^2})$ which are absolutely continuous with respect to Lebesgue measure. For every test Φ there exists a Borel set A of $\mathbb{R}^2$ such that*

$$P_s(A|X) = E_{P_s}(\Phi|X), \qquad P_s(A|Y) = E_{P_s}(\Phi|Y), \qquad \forall s = 1, \ldots, N,$$

where X and Y are the projections $(x, y) \to x$ and $(x, y) \to y$, respectively.

As in the Liapunov Theorem, this theorem follows from Theorem 1, if we can show that the orthogonal set of the linear subspace defined by the bounded statistics Z such that

$$E_{P_s}(Z|X) = E_{P_s}(Z|Y) = 0, \qquad s = 1, \ldots, N,$$

is a thin set. It is difficult to prove this property. The proof (Linnik [41, pp. 200–203]) is based on the following:

Lemma *Let A be a Borel set of $\mathbb{R}^2$ having nonzero Lebesgue measure and let k be a positive integer. Then there exist a real δ ($\delta > 0$) and a subset C having nonzero Lebesgue measure, such that the sets*

$$C_{ij} = C + (i\delta, j\delta) = \{(x + i\delta, y + j\delta),\ (x, y) \in C\}, \qquad i, j = 1, \ldots, k,$$

are disjoint and are contained in A.

An outline of the proof of Theorem 3 is as follows. First show that one can assume that the probability density functions $f_1, \ldots, f_s$ are bounded and strictly positive on A. Let C_{ij} be the sets obtained by applying the lemma with $k > 2N$. Then prove that there exists at least one measurable bounded function f which vanishes on the complement of the union of C_{ij} and for which

$$\forall (x, y) \in C; \quad i, j = 1, \ldots, k; \qquad s = 1, \ldots, N,$$

$$\begin{cases} \displaystyle\sum_{j=1}^{k} f(x + i\delta, y + j\delta) f_s(x + i\delta, y + j\delta) = 0, \\ \displaystyle\sum_{i=1}^{k} f(x + i\delta, y + j\delta) f_s(x + i\delta, y + j\delta) = 0. \end{cases}$$

Now one can verify easily that the function f satisfies

$$\forall x, y, \qquad \int_{-\infty}^{+\infty} f(\zeta, y) f_s(\zeta, y)\, d\zeta = \int_{-\infty}^{+\infty} f(x, \eta) f_s(x, \eta)\, d\eta = 0.$$

The proof is completed since for every Borel set A having nonzero Lebesgue measure, a nonzero bounded measurable function has been found which vanishes on $\complement A$ and satisfies the required condition. ∎

5. The Converse of the Neyman–Pearson Lemma

The following theorem generalizes the Hahn–Banach Theorem. It was proved first by Dubovitskii and Milyutin (for a proof see, e.g., Laurent [45, p. 24]). This theorem is useful in establishing necessary conditions for an optimum [45].

Theorem 1 *Let $U_1, \ldots, U_{p-1}$ be $p-1$ nonempty convex open subsets of a topological linear space $\mathscr{X}$ and let U_p be a nonempty convex subset of $\mathscr{X}$. Then the intersection $\bigcap_{i=1}^{p} U_i$ is empty if and only if there exist a constant c and p continuous linear forms ω_i which are not all constant, such that*

$$c \leqslant 0; \qquad \sum_{i=1}^{p} \omega_i = c; \qquad x \in U_i \;\Rightarrow\; \omega_i(x) \geqslant 0, \quad i = 1, \ldots, p.$$

Theorem 1 will now be used to prove the converse of the Neyman–Pearson Lemma,

Theorem 2 *Let μ be a positive measure on the measurable space $(\Omega, \mathscr{A})$ and let $f_0, f_1, \ldots, f_k$ be $k+1$ real measurable functions on $(\Omega, \mathscr{A})$ which are μ-integrable and linearly independent. Let Φ be a measurable function taking values in $[0, 1]$ such that, for every measurable function Φ' taking values in $[0, 1]$,*

$$\int_\Omega \Phi' f_j \, d\mu \leqslant \int \Phi f_j \, d\mu, \qquad \forall j = 1, \ldots, k \;\Rightarrow\; \int_\Omega \Phi' f_0 \, d\mu \leqslant \int_\Omega \Phi f_0 \, d\mu. \quad (1)$$

Then there exist nonnegative constants $a_0, \ldots, a_k$, such that

$$\begin{cases} a_0 f_0(\omega) > \sum_1^k f_j(\omega) a_j \;\Rightarrow\; \Phi = 1, & \mu\text{-a.e.} \\ a_0 f_0(\omega) < \sum_1^k f_j(\omega) a_j \;\Rightarrow\; \Phi = 0, & \mu\text{-a.e.} \end{cases}$$

Proof Let D be the locus in $\mathbb{R}^{k+1}$ of the points

$$\boldsymbol{\beta}_{\Phi'} = \left(\beta_{\Phi'}^0 = \int_\Omega \Phi' f_0 \, d\mu, \ldots, \beta_{\Phi'}^k = \int_\Omega \Phi' f_k \, d\mu \right),$$

where Φ' runs through all possible measurable functions having values on $[0, 1]$. This set D is a closed and convex set having a nonempty interior set. Now let $x = (x_0, \ldots, x_k)$ be a point of $\mathbb{R}^{k+1}$ and let

(a) K_j ($j = 1, \ldots, k$) be the half-spaces of $\mathbb{R}^{k+1}$ defined by $x_j < 0$,

(b) K_0 be the half-space of $\mathbb{R}^{k+1}$ defined by $x_0 > 0$,

(c) K_{k+1} be the cone of $\mathbb{R}^{k+1}$ defined by the points x, where there exists a positive real number λ, such that

$$\boldsymbol{\beta}_{\Phi} + \lambda x \in D.$$

It follows from (1) that the intersection of all these cones is empty and that they are all open and nonempty. Then Theorem 1 implies that there exist linear forms $g_0, \ldots, g_{k+1}$, not all identically zero, whose sum is zero, such that the form g_j is nonnegative on the cone K_j. That is, there exist nonnegative constants $a_0, \ldots, a_k$ which are not all zero, such that the linear form,

$$a_0(\beta_{\Phi}^0 - x_0) + \sum_1^k a_j(x_j - \beta_{\Phi}^j)$$

is nonnegative on D. Now use the same argument as in the proof of Theorem 5.4.1 to complete the proof. ∎

6. A Theorem of Linnik

The following theorem was proved by Linnik in [41]. The reader can find in Hormander [28] some results on complex analysis in several variables, needed for a more complete understanding.

Theorem 1 *Let*

$$\left[\mathbb{R}^s, \mathscr{B}_{\mathbb{R}^s}; \frac{dP_\theta}{dx} = \frac{e^{\langle\theta|x\rangle}h(x)}{l_h(-\theta)}, \quad \theta \in \Theta\right]$$

be a statistical space where Θ *is a subset of* D_h' *having a nonempty interior set. Assume that there exist an integer* s_1 *and* r *functions* $\pi_1 \cdots \pi_r$ *on* Θ, *such that if* $\mathscr{T}$ *denotes the following cone of* $\mathbb{R}^s$,

$$x_j > 0, \qquad \forall j = 1, \ldots, s_1,$$

(if $s_1 = 0$ *we choose* $\mathscr{T} = \mathbb{R}^s$*) and* V *denotes the subset of* Θ *defined by*

$$\pi_j(\theta) = 0, \qquad \forall j = 1, \ldots, r,$$

then the following conditions are satisfied:

(1) *The function* h *vanishes in* $\complement\mathscr{T}$ *and is strictly positive on* $\mathscr{T}$.

(2) Θ *contains the set* Δ

$$\theta \in (\theta_1, \ldots, \theta_s) \in \Delta \quad \Leftrightarrow \quad 0 \leqslant \theta_j \leqslant b_j, \qquad \forall j = 1, \ldots, s,$$

where the constants b_j are bounded and strictly positive. Moreover, this set Δ has a nonempty intersection with V.

(3) *There exist analytic functions $\pi_1^*, \ldots, \pi_r^*$ on the subset B of $\mathbb{C}^s$ which is defined by $\mathrm{Re}(z) \in \Delta$. Moreover, when z is real, these functions are equal to the functions $\pi_1, \ldots, \pi_r$.*

(4) *The subset V^* of B, defined by*

$$\pi_j^*(z) = 0, \qquad \forall j = 1, \ldots, r,$$

is such that, if a function is analytic on B and vanishes on V, the function vanishes on V^*.

(5) *The rank of the matrix whose general term is $\partial \pi_j^* / \partial z_i$ is equal to r for every point of B.*

(6) *There exists a ν in $\mathbb{N}^s$ and functions H_j on $\mathbb{R}^s$ which vanish in $\complement \mathbb{R}_+^s$, such that*

$$D_{H_j} \supset \Delta, \qquad l_{H_j}(z) = \pi_j^*(z) z^{-\nu}, \qquad \forall z \in B,$$

and the functions l_{H_j}, being functions of

$$(1/(1 + z_1), \ldots, 1/(1 + z_s)),$$

are analytic at the origin.

Let X be an integrable statistic whose image vanishes on $\Theta_0 = \Theta \cap V$. Then for every $\varepsilon > 0$ there exists a positive statistic δ_ε and statistics $Y_1, \ldots, Y_r$ which have derivatives of all orders and are such that†

(a) $$Xh * \delta_\varepsilon = \sum_{j=1}^{r} H_j * Y_j,$$

(b) *the statistic Y, defined by*

$$Y = \begin{cases} (Xh * \delta_\varepsilon)/h & \text{on } \mathcal{T}, \\ 0 & \text{on } \complement \mathcal{T}, \end{cases}$$

is integrable and

$$|\beta_X(\theta) - \beta_Y(\theta)| < \varepsilon, \qquad \forall \theta \in \Delta,$$

(c) $$|Y_j(x)| \leqslant K_j e^{\langle \theta | x \rangle}, \qquad \forall \theta \in \Delta, \quad \forall j = 1, \ldots, r,$$

where the K_j are constants.

Conversely, let $Y_1, \ldots, Y_r$ be r statistics, which have derivatives of all orders, satisfy (c) *and vanish in $\complement \mathcal{T}$. If X is a statistic for which there exists a δ_ε satisfying* (a), *then X is defined, is integrable and its image vanishes on Θ_0.*

When there is only one polynomial constraint, a theorem of Palamodov [55] gives an exact formula for statistics, having zero image on Θ_0.

† Note that these *statistics* are being treated as *functions*. See p. 206 for definition of *convolution* of two *functions* (Ed.).

We shall give the basic ideas behind the proof of this theorem, which is proved in detail in [41]. This proof is based on Theorem 1.3 and on the following

Lemma *Let B be the subset of $\mathbb{C}^s$, defined by*

$$0 < \operatorname{Re}(z_i) < b_i, \qquad \forall i = 1, \ldots, s.$$

Let $f_1, \ldots, f_r$ $(r < s)$ be r analytic functions on B and let W be the subset of B, defined by

$$f_j(z) = 0, \qquad \forall j = 1, \ldots, r.$$

If at every point of B the rank of the matrix $\{\partial f_j/\partial z_i\}$ is r and if the functions $f_1, \ldots, f_r$ considered as functions of $(1/(1 + z_1), \ldots, 1/(1 + z_s))$ are analytic at the origin, then every bounded analytic function G which vanishes on W can be written as

$$G = \sum_1^r G_j f_j,$$

where the functions G_j are analytic on B and satisfy

$$|G_j(z)| \leqslant k[1 + |z_1| + \cdots + |z_s|]^q, \qquad \forall z \in B', \quad j = 1, \ldots, s,$$

where K and q are positive constants and B' is the subset of B defined by

$$0 < a_i \leqslant \operatorname{Re}(z_i) \leqslant b_i' < b_i, \qquad \forall i = 1, \ldots, s.$$

Proof of Theorem 1. Let X be an integrable statistic. We have

$$D_{Xh} \supset D_h$$

and then the function,

$$\varphi_X(z) = l_{Xh}(z) = \int_{\mathbb{R}^s} e^{-\langle z|x\rangle} X(x)h(x)\,dx = \int_{\mathscr{T}} e^{-\langle z|x\rangle} Xh\,dx$$

is defined if $\operatorname{Re}(z) \in D_h$. Moreover,

$$|\varphi_X(z)| \leqslant \int_{\mathscr{T}} e^{-\langle \operatorname{Re}(z)|x\rangle} |X| h\,dx = \varphi_{|X|}(\operatorname{Re}(z)).$$

If z is real and equal to θ, $\varphi_{|X|}(\theta)$ is continuous and thus bounded on the compact set Δ. Hence $\varphi_X(z)$ is bounded on B.

If now X has a zero image β_X on Θ_0, using Condition (2), we see that β_X vanishes on $\Delta \cap V$. Since

$$\beta_X(\theta) = \frac{\varphi_X(\theta)}{l_h(-\theta)},$$

$\varphi_X(z)$ vanishes on V. Therefore Condition 4, implies that $\varphi_X(z)$ also vanishes on V^*. Using Assumptions (3), (5) and (6) we can apply the lemma with

$$W = V^*, \qquad f_j = l_{H_j}.$$

Then there exist functions G which are analytic on B, such that

$$\begin{cases} \varphi_X(z) = \sum_1^r l_{H_j}(z) G_j(z), & \forall z \in B \\ |G_j(z)| \leqslant K[1 + |z_1| + \cdots + |z_s|]^q, & \forall z \in B'. \end{cases}$$

Let δ_ε be a positive function whose integral equals 1, which has derivatives of all orders and is zero except in the subset $[0, \eta]^s$, where $\eta = \eta(\varepsilon)$ will be chosen later. Since

$$D_{Xh} \supset D_h, \qquad D_{\delta_\varepsilon} = \mathbb{R}^s,$$

the convolution of Xh and δ_ε is well defined and vanishes on $\complement\mathcal{T}$ because Xh and δ_ε vanish on $\complement\mathcal{T}$. Then the statistic Y, satisfying

$$Yh = Xh * \delta_\varepsilon,$$

is well-defined, and

$$\varphi_Y(z) = \varphi_X(z) l_{\delta_\varepsilon}(z), \qquad \forall z \in B.$$

Let

$$G_j(z) l_{\delta_\varepsilon}(z) = K_j(z), \qquad \forall j = 1, \ldots, r.$$

Then

$$\varphi_Y(z) = \sum_1^r l_{H_j}(z) K_j(z).$$

But δ_ε has derivatives of all orders and, for all integral k,

$$|l_{\delta_\varepsilon}(z)| \leqslant K'[(1 + |z_1|) \cdots (1 + |z_s|)]^{-k}, \qquad \forall z \in \mathbb{C}^s.$$

Then, for all positive m,

$$|K_j(z)| \leqslant K''[(1 + |z_1|) \cdots (1 + |z_s|)]^{-m}, \qquad \forall z \in B'.$$

Using Theorem 1.3, we see that the function $K_j(z)$ is the Laplace transform of a statistic Y_j, satisfying (c), such that

$$D_{Y_j} \supset \Delta.$$

It is easy to see that Y_j has derivatives of any order. Now, we use Assumption (6) to define the convolution $Y_j * H_j$ and then (a) is proved, since statistics are functions on $\mathbb{R}^s$.

On the other hand, we have

$$\beta_Y(\theta) - \beta_X(\theta) = \beta_X(\theta)[1 - l_{\delta_\varepsilon}(\theta)],$$

and β_X is bounded on the compact set Δ. Since the integral of δ_ε equals 1, we have

$$|1 - l_{\delta_\varepsilon}(\theta)| \leqslant 1 - e^{-\eta\theta}.$$

Now, we can choose η such that

$$|1 - l_{\delta_\varepsilon}(\theta)| \leqslant \varepsilon$$

and (b) is proved.

Conversely, we have

$$\varphi_X(z) l_{\delta_\varepsilon}(z) = \sum_1^r l_{H_j}(z) K_j(z),$$

and then $\varphi_X(z) l_{\delta_\varepsilon}(z)$ vanishes on V. But $l_{\delta_\varepsilon}(z)$ is positive when z is real. Hence $\varphi_X(\theta)$ vanishes on Θ_0. ∎

Exercises

1. Consider the statistical space,

$$[\mathbb{R}, \mathscr{B}_{\mathbb{R}}; N(m, \sigma^2), (m, \sigma^2) \in \Theta \subset \mathbb{R} \times \mathbb{R}^+]^n.$$

(a) Find a $\mathscr{P}$-minimum complete sufficient statistic when $n > 2$ and the interior set of Θ is nonempty.

(b) Show that there exists a scalar-valued continuously differentiable sufficient statistic if and only if $\dim \Theta = 2$.

2. For every positive ν belonging to $\mathbb{N}^s$ (notation given in beginning of Section 10.1), compute the Laplace transform of the function,

$$x \in (\mathbb{R}^+)^s \to x^\nu/\nu!.$$

3. Consider the statistical space,

$$(\mathbb{R}^n, \mathscr{B}_{\mathbb{R}^n}, \mathscr{P}).$$

Assume that it is dominated by Lebesgue measure and that there exists a distribution $P_0 \in \mathscr{P}$, whose probability density function is strictly positive on a given subset Ω_0. Show that this subset Ω_0 satisfies the condition of Definition 2.1.

4. Show that $U \cap f^{-1}(\mathscr{B}_{\mathbb{R}^n})$ is the σ-field of the Borel sets of U for the U of Theorem 2.2.

5. Consider Theorem 2.3 and assume that $r = n$. Show that

$$U \cap f^{-1}(\mathscr{B}_{\mathbb{R}^m}) = \mathscr{B}_U,$$

where $\mathscr{B}_U$ is the σ-field of the Borel sets of U.

6. Prove Theorem 2.5. (*Hint:* First show that one can restrict oneself to the case $n = k$. Then, for this case, prove the theorem by induction on k; verify it for $k = 1$ and prove by *reductio ad absurdum*, that it is true for k if it is true for $k - 1$.)

7. What does Theorem 2.7 mean when the dimension of $\mathscr{L}$ is infinite? Under the assumptions of this theorem, show that if there exists some integer n, such that the statistical space

$$(I, \mathscr{B}_I; p_\theta, \theta \in \Theta)^n$$

has a continuously differentiable sufficient statistic whose dimension is less than n, then the statistical space $(I, \mathscr{B}_I; p_\theta, \theta \in \Theta)$ is an exponential one.

Consider a sample of size n such that for every n greater than a given n_0 there exists a sufficient statistic whose dimension is less than n. Find general assumptions under which the dimension of this sufficient statistic does not depend on n.

8. Converse of Theorem 4.2. Using notations of Definition 4.1, show that if a finite set W is thin, then P^* is nonatomic.

9. Using notations of the Linnik Theorem (Theorem 6.1), show that if the functions π_j are polynomials then Conditions 3 and 6 are realized.

10. Let f be a function on $\mathbb{R}^s$, having derivatives of order r and let $\nu \in \mathbb{N}^s$, $|\nu| \leqslant r$. (See Section 10.1.) Give conditions on f, for the Laplace transform of

$$g(x) = \frac{\partial^{|\nu|}}{\partial x^\nu} f(x)$$

to be equal to $z^\nu l_f(z)$. Find a converse of this result. [Using notation of Section 10.1, the differential operator $\partial^{|\nu|}/\partial x^\nu$, where both $x = (x_1, x_2, \ldots, x_s)$ and ν are vectors, is defined as

$$\frac{\partial^{|\nu|}}{\partial x^\nu} = \frac{\partial^{\Sigma n_i}}{\prod_{i=1}^s \partial x_i^{n_i}}.\Bigg]$$

11. Let I be an interval of $\mathbb{R}$, let $\mathscr{B}_I$ be the σ-field of Borel sets of I and let $(I, \mathscr{B}_I; p_\theta, \theta \in \Theta)$ be a statistical space dominated by Lebesgue measure on I. Assume that there exists an integer $n > 1$ such that the statistical space $(I, \mathscr{B}_I; p_\theta, \theta \in \Theta)^n$ admits a differentiable real-valued sufficient statistic. Show *directly* that if p_θ is strictly positive and differentiable on I then the linear space generated by the functions $\log(p_\theta(x)/p_{\theta_0}(x))$ has dimension 1.

12. Show that if there exists a one-to-one continuously differentiable mapping from an open set A of $\mathbb{R}^k$ into $\mathbb{R}^n$, then $n \geqslant k$.

13. How can the Linnik Theorem be simplified when $r = 1$?

14. Consider the following statistical space

$$[\Omega, \mathscr{A}; P_\theta, \theta \in (\Theta, \mathscr{T})]$$

and let m be a bounded measure on $(\Theta, \mathscr{T})$. Show that there exists a measure μ on $(\Omega, \mathscr{A})$ such that, for every integrable statistic X whose image β_X is m-integrable,

$$\int_\Theta \beta_X \, dm = \int_\Omega X \, d\mu.$$

What is the relationship between the mapping β and the mapping $m \to \mu$?

15. Generalize Theorem 4.3 to the case of distributions on $(\mathbb{R}^n, \mathscr{B}_{\mathbb{R}^n})$ instead of $(\mathbb{R}^2, \mathscr{B}_{\mathbb{R}^2})$.

16. For a statistical space on a finite set, state and prove a theorem, analogous to Linnik's Theorem.

APPENDIX

Conditional Probability

In this book we have often used properties of conditional probabilities. We now recall these properties with abbreviated proofs or references. In some cases, the proofs are left as exercises for the reader. We designate by $\overline{\mathbb{R}}$ the space $\mathbb{R}$ plus the points at infinity. Unless stated otherwise, throughout the chapter all random elements are defined on the same basic probability space $(\Omega, \mathscr{A}, P)$, $\mathscr{B}$ is a subfield of $\mathscr{A}$ and T is a random element taking values on $(\mathscr{X}, \mathscr{C})$.

1. Preliminary Results

Lemma 1 [53, p. 36; 52, p. 243] *Let* $(\Omega, \mathscr{A})$ *be a measurable space, let* X *be a measurable mapping from* $(\Omega, \mathscr{A})$ *into* $(\mathscr{X}, \mathscr{C})$ *and let* Y *be a measurable mapping from* $(\Omega, \mathscr{A})$ *into* $(\mathbb{R}, \mathscr{B}_{\mathbb{R}})$ *(or into* $(\overline{\mathbb{R}}, \mathscr{B}_{\overline{\mathbb{R}}})$*). Then* Y *is* $X^{-1}(\mathscr{C})$*-measurable if and only if there exists a measurable mapping* g *from* $(\mathscr{X}, \mathscr{C})$ *into* $(\mathbb{R}, \mathscr{B}_{\mathbb{R}})$ *(or into* $(\overline{\mathbb{R}}, \mathscr{B}_{\overline{\mathbb{R}}})$*) such that*

$$Y = g(X).$$

Theorem 1 [52, *p.* 86] *Let* $(\Omega, \mathscr{A})$ *be a measurable space, let* T *be a measurable mapping from* $(\Omega, \mathscr{A})$ *into* $(\mathscr{X}, \mathscr{C})$*, let* μ *be a measure on* $(\Omega, \mathscr{A})$ *and let* μ_T *be the measure induced by* T *on* $(\mathscr{X}, \mathscr{C})$*. A real-valued measurable* *function g defined on $(\mathscr{X}, \mathscr{C})$ is μ_T-integrable if and only if $g(T)$ is μ-integrable and then we have*

$$\forall C \in \mathscr{C}, \qquad \int_{T^{-1}(C)} g(T)\, d\mu = \int_C g\, d\mu_T.$$

Theorem 2 [52, *pp.* 80 *and* 261] *Let μ be a positive measure on $(\Omega, \mathscr{A})$. For every positive measurable (or μ-integrable) function f the mapping*

$$A \in \mathscr{A} \rightarrow m(A) = \int_A f\, d\mu$$

defines a positive (or real-valued measure) m on $(\Omega, \mathscr{A})$ which is absolutely continuous with respect to μ, i.e.,

$$A \in \mathscr{A}, \qquad \mu(A) = 0 \;\Rightarrow\; m(A) = 0.$$

When m is a probability distribution, we say that f is a determination of the probability density function of m with respect to μ.

Theorem 3 [52, *p.* 82] *A real-valued measurable function h is m-integrable, if and only if $f \cdot h$ is μ-integrable, the measures m and μ being given in Theorem 2. Then we have*

$$\forall A \in \mathscr{A}, \qquad \int_A h\, dm = \int_A h \cdot f\, d\mu.$$

We say that the positive measure μ on $(\Omega, \mathscr{A})$ is σ-finite if Ω is a countable union of subsets Ω_n such that

$$\Omega_n \in \mathscr{A}, \qquad \mu(\Omega_n) < \infty.$$

Theorem 4 (Radon–Nikodym) [53, p. 111; 52, p. 267] *Let μ be a positive and σ-finite measure on $(\Omega, \mathscr{A})$ and let m be a positive measure which is absolutely continuous with respect to μ. Then there exists a positive measurable real-valued function f such that*

$$\forall A \in \mathscr{A}, \qquad m(A) = \int_A f\, d\mu. \tag{1}$$

This function f is unique μ-a.e. Moreover, f is bounded (μ-a.e.) if and only if m is σ-finite and f is μ-integrable if and only if m is bounded.

We shall denote by $dm/d\mu$ the family of all functions which satisfy (1) and are all equal μ-a.e. Using Theorem 3 we see that

$$\frac{dm}{d\mu} = \left\{ hg : h \in \frac{dm}{d\nu}, \quad g \in \frac{d\nu}{d\mu} \right\}.$$

When there is no ambiguity, we shall write

$$f = \frac{dm}{d\mu} \qquad \text{and} \qquad \frac{dm}{d\mu} = \frac{dm}{d\nu}\frac{d\nu}{d\mu}.$$

Remark When μ is a probability distribution [53, p. 111] or when m is σ-finite [62, pp. 98–100], this theorem remains valid even when the assumption that m is positive is omitted.

2. Conditional Expectation (with Respect to a σ-Field or a Random Element)

Theorem 1 *Let X be a P-equivalence class of random elements (see Section 12.3) taking values in $(\overline{\mathbb{R}}^+, \mathscr{B}_{\overline{\mathbb{R}}^+})$ and let $\mathscr{B}$ be a subfield of $\mathscr{A}$. Then there exists a unique P-equivalence class, designated by $E^*(X|\mathscr{B})$, of $\mathscr{B}$-measurable random elements taking values in $(\overline{\mathbb{R}}^+, \mathscr{B}_{\overline{\mathbb{R}}^+})$, such that*

$$\forall B \in \mathscr{B}, \qquad \forall Y \in E^*(X|\mathscr{B}), \qquad \int_B^* X\, dP = \int_B^* Y\, dP.$$

Here we have designated by $\int^*$ the upper integral, which is also the expectation in the sense of [53, pp. 37–39].

Proof The P-equivalence class X defines a unique measure ν_X on $(\Omega, \mathscr{A})$ defined by

$$\forall A \in \mathscr{A}, \qquad \nu_X(A) = \int_A X\, dP.$$

The measure ν_X is positive and absolutely continuous with respect to P (Theorem 1.2). Let ν and μ be the restrictions of ν_X and P on $(\Omega, \mathscr{B})$.

Apply the Radon–Nikodym theorem and designate by $E^*(X|\mathscr{B})$ the unique equivalence class $d\nu/d\mu$ thus obtained. Then verify that

$$Y \in E^*(X|\mathscr{B}) \quad \Leftrightarrow \quad Y = \frac{d\nu}{d\mu}$$

$$\Leftrightarrow \quad \forall B \in \mathscr{B}, \quad \nu(B) = \int_B Y\, d\mu = \int_B Y\, dP = \int_B X\, dP. \qquad \blacksquare$$

Theorem 2 *Let X be a P-equivalence class of integrable random variables and let $\mathscr{B}$ be a subfield of $\mathscr{A}$. Then there exists a unique P-equivalence class, designated by $E(X|\mathscr{B})$, of $\mathscr{B}$-measurable integrable random variables, such that*

$$\forall B \in \mathscr{B}, \qquad \forall Y \in E(X|\mathscr{B}), \qquad E(1_B X) = E(1_B Y).$$

Proof Let X^+ and X^- be the positive and the negative parts of X, i.e.,

$$X = X^+ - X^-, \qquad X^+ = \sup(X, 0), \qquad X^- = \sup(-X, 0).$$

Since the measures ν_{X^-} and ν_{X^+} are bounded (because X^+ and X^- are integrable), then by the Radon–Nikodym Theorem, the functions belonging to $E^*(X^+ | \mathscr{B})$ and $E^*(X^- | \mathscr{B})$ are integrable. Then there exists nonempty equivalence classes $E(X^+ | \mathscr{B})$ and $E(X^- | \mathscr{B})$ which are included in $E^*(X^+ | \mathscr{B})$ and $E^*(X^- | \mathscr{B})$, respectively. The equivalence class $E(X | \mathscr{B})$ defined by

$$E(X | \mathscr{B}) = E(X^+ | \mathscr{B}) - E(X^- | \mathscr{B}),$$

satisfies the conditions of the theorem. As an exercise, the reader should show that $E(X | \mathscr{B})$ is unique.

Let T be a random element taking values in $(\mathscr{X}, \mathscr{C})$. Then the σ-field $T^{-1}(\mathscr{C})$ is a subfield of $\mathscr{A}$ and we define $E(X | T)$ or $E^*(X | T)$ by $E(X | T^{-1}(\mathscr{C}))$ or $E(X^* T^{-1}(\mathscr{C}))$ according as X is integrable or positive, respectively. These equivalence classes are called conditional expectations of X given T and when there is no ambiguity we identify such a class with an arbitrary element of that class.

Theorem 3 *Let X be an integrable random variable and let T be a random element taking values in $(\mathscr{X}, \mathscr{C})$. Then there exists a unique equivalence class of integrable random variables which depends only on T, such that*

$$\forall C \in \mathscr{C}, \qquad \forall Y \in E(X | T), \qquad \int_{T^{-1}(C)} X \, dP = E(1_C(T) \cdot Y).$$

Proof Using Theorem 2 we have

$$\forall C \in \mathscr{C}, \qquad \forall Y \in E(X | T^{-1}(\mathscr{C})) = E(X | T), \qquad \int_{T^{-1}(C)} X \, dP = \int_{T^{-1}(C)} Y \, dP.$$

Since Y is $T^{-1}(\mathscr{C})$-measurable Lemma 1.1 shows that there exists a measurable mapping g from $(\mathscr{X}, \mathscr{C})$ into $(\mathbb{R}, \mathscr{B}_{\mathbb{R}})$ such that

$$Y = g(T).$$

Therefore,

$$\int_{T^{-1}(C)} X \, dP = \int_{T^{-1}(C)} g(T) \, dP.$$

Applying Theorem 1.1 we have

$$\int_{T^{-1}(C)} X \, dP = \int_C g \, dP_T. \qquad \blacksquare$$

3. Conditional Probability (with Respect to a σ-Field or a Random Element)

We specialize the previous concepts in the case of random variables which are set indicators, the reader should not confuse conditional probabilities with conditional distributions defined in Section 5.

Theorem 1 *Let $A \in \mathscr{A}$ be an event and let $\mathscr{B}$ be a subfield of $\mathscr{A}$. Then there exists a unique P-equivalence class $P(A|\mathscr{B})$ of $\mathscr{B}$-measurable integrable random variables, such that*

$$\forall B \in \mathscr{B}, \quad \forall Y \in P(A|\mathscr{B}), \qquad P(A \cap B) = \int_B Y\, dP.$$

Proof Applying Theorem 2.2 to the random variable 1_A, we obtain

$$P(A|\mathscr{B}) = E(1_A|\mathscr{B}). \qquad \blacksquare$$

Theorem 2 *Let $A \in \mathscr{A}$ be an event and let T be a random element taking values in $(\mathscr{X}, \mathscr{C})$. Then there exists a unique P-equivalence class $P(A|T)$ of integrable random variables depending only on T, such that*

$$\forall C \in \mathscr{C}, \quad \forall Y \in P(A|T), \qquad P(A \cap T^{-1}(C)) = E(1_C(T)Y).$$

We call $P(A|T)$ the *conditional probability of A given T*.

Remark 1 Let B be an event such that $0 < P(B) < 1$ and let $\mathscr{B}_B$ be the σ-field $\{\varnothing, \Omega, B, \complement B\}$. Then

$$\frac{P(A \cap B)}{P(B)} 1_B + \frac{P(A \cap \complement B)}{P(\complement B)} 1_{\complement B}$$

is a determination of $P(A|\mathscr{B}_B)$.

Remark 2 $P(A|\mathscr{B})$ (or $P(A|T)$) is called the conditional probability of A given $\mathscr{B}$ (or T).

4. Properties of Conditional Expectation

Theorem 1 (Nonnegativity) *Let X be a random element taking values in $(\overline{\mathbb{R}}, \mathscr{B}_{\overline{\mathbb{R}}})$ (or an integrable random variable), then*

$$X \geqslant 0, \quad P\text{-a.e.} \Rightarrow E^*(X|\mathscr{B}) \geqslant 0 \quad (\text{or } E(X|\mathscr{B}) \geqslant 0) \quad P\text{-a.e.}$$

Theorem 2 (Linearity) *Let X and Y be two random elements taking values in $(\overline{\mathbb{R}}^+, \mathscr{B}_{\overline{\mathbb{R}}^+})$ (or integrable random variables), then*

$$E^*(X + Y|\mathscr{B}) = E^*(X|\mathscr{B}) + E^*(Y|\mathscr{B})$$
$$(\text{or } E(X|\mathscr{B}) + E(Y|\mathscr{B}) = E(X + Y|\mathscr{B}))$$

$$c \in \mathbb{R}^+, \qquad E^*(cX|\mathscr{B}) = cE^*(X|\mathscr{B})$$
$$(\text{or } c \in \mathbb{R}, \quad E(cX|\mathscr{B}) = cE(X|\mathscr{B})).$$

Theorem 3 (Monotonicity) *Let X and Y be two random elements taking values on $(\overline{\mathbb{R}}^+, \mathscr{B}_{\overline{\mathbb{R}}^+})$ and let X' and Y' be two integrable random variables. Then*

$$X \leqslant Y, \quad P\text{-a.e.} \Rightarrow E^*(X|\mathscr{B}) \leqslant E^*(Y|\mathscr{B}), \quad P\text{-a.e.} \tag{1}$$

$$X' \leqslant Y', \quad P\text{-a.e.} \Rightarrow E(X'|\mathscr{B}) \leqslant E(Y'|\mathscr{B}), \quad P\text{-a.e.} \tag{2}$$

$$X' \leqslant Y, \quad P\text{-a.e.} \Rightarrow E(X'|\mathscr{B}) \leqslant E^*(Y|\mathscr{B}), \quad P\text{-a.e.} \tag{3}$$

$$X \leqslant Y', \quad P\text{-a.e.} \Rightarrow E^*(X|\mathscr{B}) \leqslant E(Y'|\mathscr{B}), \quad P\text{-a.e.} \tag{4}$$

Furthermore, if Z is a $\mathscr{B}$-measurable random element taking values in $(\overline{\mathbb{R}}, \mathscr{B}_{\overline{\mathbb{R}}})$, then

$$X' \leqslant Z, \quad P\text{-a.e.} \Rightarrow E(X'|\mathscr{B}) \leqslant Z, \quad P\text{-a.e.} \tag{5}$$

Proof Implications (1) and (2) are deduced directly from Theorems 1 and 2. Properties (3) and (4) are proved by separating the positive and negative parts of the random elements which are involved. For example, (3) is derived as follows:

$$X' \leqslant Y, \quad P\text{-a.e.} \Leftrightarrow X'^+ \leqslant Y + X'^-. \quad P\text{-a.e.},$$

Then

$$E^*(X'^+|\mathscr{B}) \leqslant E^*(Y + X'^-|\mathscr{B}), \quad P\text{-a.e.},$$

$$E^*(X'^+|\mathscr{B}) = E(X'^+|\mathscr{B}), \quad P\text{-a.e.},$$

$$E^*(Y + X'^-|\mathscr{B}) = E^*(Y|\mathscr{B}) + E^*(X'^-|\mathscr{B}), \quad P\text{-a.e.},$$

$$E^*(X'^-|\mathscr{B}) = E(X'^-|\mathscr{B}), \quad P\text{-a.e.},$$

whence

$$E(X'^+|\mathscr{B}) \leqslant E^*(Y|\mathscr{B}) + E(X'^-|\mathscr{B}), \quad P\text{-a.e.}$$

Thus

$$E(X'^+|\mathscr{B}) - E(X'^-|\mathscr{B}) = E(X'|\mathscr{B}) \leqslant E^*(Y|\mathscr{B}). \quad P\text{-a.e.}$$

To prove (5), we let B be the subset

$$B = \{E(X'|\mathscr{B}) > Z\}.$$

It belongs to $\mathscr{B}$ and by assumption we have

$$\int_B [X' - E(X'|\mathscr{B})]\, dP = 0.$$

But on B,

$$X' < E(X'|\mathscr{B}),$$

and then

$$P(B) = 0.$$

∎

Theorem 4 (Integrability) *Let X be a random element taking values in $(\overline{\mathbb{R}}, \mathscr{B}_{\overline{\mathbb{R}}})$, then X is integrable if and only if $E^*(|X|\,|\mathscr{B})$ is integrable. Thus we have*

$$E(X) = E[E(X|\mathscr{B})], \qquad |E(X|\mathscr{B})| \leqslant E(|X|\,|\mathscr{B}).$$

Proof If X is positive, this property is proved by using the Radon–Nikodym Theorem and the proof of Theorem 2.1. If X is integrable, then X^+ and X^- are also integrable (and conversely) and

$$E^*(X^+|\mathscr{B}) + E^*(X^-|\mathscr{B}) = E^*(|X|\,|\mathscr{B})$$

is also integrable. Applying Theorem 2.2 with $B = \Omega$, we obtain

$$E(X) = E[E(X|\mathscr{B})].$$

Then

$$E(X|\mathscr{B}) = E(X^+|\mathscr{B}) - E(X^-|\mathscr{B}),$$

$$|E(X|\mathscr{B})| \leqslant E(X^+|\mathscr{B}) + E(X^-|\mathscr{B}) = E(|X|\,|\mathscr{B}).$$

∎

Theorem 5 (Monotone continuity) *Let $\{X_n\}$ be an increasing sequence of random elements taking values in $(\overline{\mathbb{R}}, \mathscr{B}_{\overline{\mathbb{R}}^+})$, which converges to X, then*

$$E^*(X|\mathscr{B}) = \lim_n \uparrow E^*(X_n|\mathscr{B}), \qquad P\text{-a.e.}$$

Moreover let $\{X_n\}$ be an increasing sequence of integrable random variables which are uniformly bounded. Then there exists an integrable random variable X, such that

$$\lim_n X_n = X, \qquad P\text{-a.e.}, \qquad E(X|\mathscr{B}) = \lim_n E(X_n|\mathscr{B}), \qquad P\text{-a.e.}$$

These monotonicity properties are also valid for decreasing sequences.

Corollary 1 *Let $\{X_n\}$ be a sequence of integrable random variables such that the series $\sum_n E(|X_n|)$ converges. Then there exists an integrable random variable X, such that*

$$X = \sum_n X_n, \qquad P\text{-a.e.}$$

$$E(X|\mathscr{B}) = \sum_n E(X_n|\mathscr{B}), \qquad P\text{-a.e.}$$

Proof Separate X_n into its positive and negative parts

$$X_n = X_n^+ - X_n^-,$$

and apply the previous property to the increasing sequences

$$U_n = \sum_1^n X_i^+, \qquad V_n = \sum_1^n X_i^-,$$

which are such that

$$E(U_n) \leqslant \sum_{n=1}^{\infty} E(|X_n|), \qquad E(V_n) \leqslant \sum_{n=1}^{\infty} E(|X_n|).$$

These sequences have as limits U and V respectively, with

$$E(U|\mathscr{B}) = \lim_{n\to\infty} \left(\sum_1^n E(X_i^+|\mathscr{B}) \right),$$

$$E(V|\mathscr{B}) = \lim_{n\to\infty} \left(\sum_1^n E(X_i^-|\mathscr{B}) \right).$$

Now choose $X = U - V$ which satisfies the conditions of the corollary, since

$$E(X|\mathscr{B}) = E(U|\mathscr{B}) - E(V|\mathscr{B}) = \lim_{n\to\infty} \left(\sum_1^n E(X_i|\mathscr{B}) \right) = \sum_{n=1}^{\infty} E(X_n|\mathscr{B}), \quad P\text{-a.e.} \quad \blacksquare$$

Corollary 2 (Fatou–Lebesgue) *Let U be a positive and integrable random variable. If*

$$|X_n| \leqslant U, \qquad \lim_n X_n = X, \qquad P\text{-a.e.}$$

then X is an integrable random variable, satisfying

$$E(X|\mathscr{B}) = \lim_n E(X_n|\mathscr{B}).$$

Theorem 6 (Transitivity) *If $\mathscr{B} \subset \mathscr{B}'$, then*

$$E(X|\mathscr{B}) = E[E(X|\mathscr{B})|\mathscr{B}'] = E[E(X|\mathscr{B}')|\mathscr{B}].$$

The following special cases are useful:

$$E(X|g(T)) = E[E(X|T)|g(T)] = E[E(X|g(T))|T];$$

$$E(X|U) = E[E(X|(U,V))|U] = E[E(X|U)|(U,V)].$$

Theorem 7. *Let T, V be two random elements where T takes values in $(\mathscr{X},\mathscr{C})$. Then V and T are independent if and only if*

$$\forall C \in \mathscr{C}, \qquad P(T^{-1}(C)|V) = P(T^{-1}(C)), \qquad P\text{-a.e.}$$

Moreover, if T is also an integrable random variable,

$$E(T|V) = E(T), \qquad P\text{-a.e.}$$

Proof Let $(\mathscr{Y},\mathscr{D})$ be the space in which V takes values. Then T and V are independent if and only if the subfield $T^{-1}(\mathscr{C})$ and $V^{-1}(\mathscr{D})$ are independent, i.e., if and only if

$$\forall C \in \mathscr{C}, \quad D \in \mathscr{D}, \qquad P(T^{-1}(C) \cap V^{-1}(D)) = P(T^{-1}(C))P(V^{-1}(D)),$$

or

$$\forall C \in \mathscr{C}, \quad D \in \mathscr{D}, \qquad P(T^{-1}(C) \cap V^{-1}(D)) = \int_{V^{-1}(D)} P(T^{-1}(C))\, dP.$$

Thus we can choose $P(T^{-1}(C))$ as a determination of $P(T^{-1}(C)|V)$. Now if T is an integrable random variable, for every real valued function g, we have

$$E(Tg(V)) = E(T)E(g(V)) = E[E(T)g(V)].$$

Finally, choose $E(T)$ for the determination of $E(T|V)$. ∎

Theorem 8 (Multiplicativeness) *Let X and Z be two random elements taking values in $(\overline{\mathbb{R}}^+, \mathscr{B}_{\mathbb{R}^+})$. If Z is $\mathscr{B}$-measurable we have*

$$E^*(XZ|\mathscr{B}) = ZE^*(X|\mathscr{B}), \qquad P\text{-a.e.}$$

Let X be an integrable random variable and let Z be a $\mathscr{B}$-measurable random variable. Then XZ is integrable if and only if $ZE(X|\mathscr{B})$ is integrable; then

$$E(XZ|\mathscr{B}) = ZE(X|\mathscr{B}), \qquad P\text{-a.e.}$$

The second part of Theorem 8 is proved by separating the positive and the negative parts of $X, Z, Y = E(X|\mathscr{B})$ and applying the first part of Theorem 8 to each of the four terms which are obtained. The following special cases are interesting:

$$E(Yf(T)|T) = f(T)E(Y|T), \qquad P\text{-a.e.}$$

$$\int_A f(T)\,dP = \int f P(A|T)\,dP_T.$$

Definition 1 Let $X = (X_1, \ldots, X_n)$ be an integrable (or positive) random vector. Then we define

$$E(X|\mathscr{B}) = (E(X_1|\mathscr{B}), \ldots, E(X_n|\mathscr{B}))$$
$$(or\ E^*(X|\mathscr{B}) = (E^*(X_1|\mathscr{B}), \ldots, E^*(X_n|\mathscr{B}))).$$

Theorem 9 (Jensen Inequality) *Let φ be a real-valued continuous and convex function defined on a convex subset K of $\mathbb{R}^n$. Let $\mathscr{B}$ be a subfield of $\mathscr{A}$ and let X be an integrable random vector, taking values in K. If $\varphi(X)$ is integrable and if there exists a determination of $E(X)|\mathscr{B})$, taking values in K, then $\varphi(E(X)|\mathscr{B})$ is integrable and*

$$E(\varphi(X)|\mathscr{B}) \geqslant \varphi(E(X|\mathscr{B})), \qquad P\text{-a.e.}$$

Proof Since φ is convex, there exists a Borel function λ on K, such that

$$\forall x, \quad x' \in K, \qquad \varphi(x) - \varphi(x') \geqslant \langle \lambda(x') | x - x' \rangle.$$

Then

$$\varphi(X) - \langle \lambda(E(X|\mathscr{B})) | X - E(X|\mathscr{B}\rangle \geqslant \varphi(E(X|\mathscr{B})),$$

and

$$\varphi(E(X|\mathscr{B})) \geqslant \varphi(E(X)) + \langle \lambda(E(X)) | E(X|\mathscr{B}) - E(X)\rangle.$$

Using Theorem 8 and taking the conditional expectation given $\mathscr{B}$ one can prove that the previous two inner products vanish. Then $|\varphi(E(X|\mathscr{B}))|$ is bounded by an integrable function and thus is integrable. Finally apply Theorem 3. ∎

The following special cases are useful:

Corollary 3 *Let X be a random variable and let p be a real number greater than 1. If $|X|^p$ is integrable, $|E(X|\mathscr{B})|^p$ is also integrable and*

$$E(|X|^p) \geqslant E(|E(X|\mathscr{B})|^p).$$

Corollary 4 *Let X be a random variable whose square is integrable. Then* $\mathbf{X} = E(X|\mathscr{B})$ *also has an integrable square and*

$$E(\mathbf{X}^2) \leqslant E(X^2), \qquad \sigma^2(\mathbf{X}) \leqslant \sigma^2(X).$$

Furthermore, for every $\mathscr{B}$-measurable random variable Y having an integrable square we have

$$E(X - Y)^2 \geqslant E(X - \mathbf{X})^2.$$

For a proof of these corollaries see [6, p. 244].

Remark Let X and Y be two random vectors taking values on the same space, each having a covariance matrix. Then each of the following two vectors are uncorrelated with Y

$$Y' = X - E(X) - \Sigma_{XY}\Lambda_Y^{-1}(Y - E(Y)),$$

$$Y'' = X - E(X|Y).$$

The previous results show that the covariance of Y'' is less than the covariance of Y'.

The following equation is called the *linear regression equation*

$$X = E(X) + \Sigma_{XY}\Lambda_Y^{-1}(Y - E(Y)) + Y'$$

and

$$X = E(X|Y) + Y''.$$

is called the *conditional regression equation.* We have seen in Section 7.4 that, in the normal case, these two equations are identical.

5. Conditional Distribution

Definition 1 Let $\mathscr{B}$ be a subfield of $\mathscr{A}$. Assume that there exist a subfield $\mathscr{A}'$ of $\mathscr{A}$ and a function $P_{\mathscr{B}}(A, \omega)$ on $\mathscr{A}' \times \Omega$, such that

(a) for every $\omega \in \Omega$, the function $P_{\mathscr{B}}^{\omega}$: $A \to P_{\mathscr{B}}(A, \omega)$, when treated as a function of A, $P_{\mathscr{B}}^{\omega}(A)$, is a probability distribution on $(\Omega, \mathscr{A}')$;
(b) for every $A \in \mathscr{A}'$, the function $\ell_{\mathscr{B}}^{A}$: $\omega \to P_{\mathscr{B}}(A, \omega)$, when treated as a function of ω, $\ell_{\mathscr{B}}^{A}(\omega)$, is a determination of $P(A|\mathscr{B})$.

Then, $P_{\mathscr{B}}^{\omega}$, the family of probability distributions when ω runs through Ω, is called the *conditional distribution on $\mathscr{A}'$ given $\mathscr{B}$*, i.e.,

$$\forall A \in \mathscr{A}', \qquad P_{\mathscr{B}}^{\omega}(A) = P_{\mathscr{B}}(A, \omega).$$

Such a conditional distribution does not always exist; see, for example, [62, p. 232] for conditions under which it exists.

Theorem 1 *If there exists a conditional distribution $P_{\mathscr{B}}^{\omega}$ on $\mathscr{A}'$ given $\mathscr{B}$, then for every $\mathscr{A}'$-measurable and integrable random variable X, the random variable*

$$\omega \to \int_{\Omega} X \, dP_{\mathscr{B}}^{\omega}$$

is defined for every ω and is a determination of $E(X|\mathscr{B})$.

Definition 2 Let X and Y be two random elements taking values in $(\mathscr{X}, \mathscr{C})$ and $(\mathscr{Y}, \mathscr{D})$, respectively with distributions P_X and P_Y. We call the *conditional distribution of Y given X*, when it exists, a function $P_{Y|X}^{x}(D)$ on $\mathscr{X} \times \mathscr{D}$ such that

(a) for every $x \in \mathscr{X}$, the function $P_{Y|X}^{x}$: $D \to P_{Y|X}^{x}(D)$, when treated as a function of D, $P_{Y|X}^{x}(D)$, is a probability distribution on $(\mathscr{Y}, \mathscr{D})$

(b) for every $D \in \mathscr{D}$, the function $h_{Y|X}^{D}$: $x \to P_{Y|X}^{x}(D)$, when treated as a function of x, is such that $h_{Y|X}^{D}(X)$ is a determination of $P(Y^{-1}(D)|X)$.

We remark that condition (b) is equivalent to

$$P(X^{-1}(C) \cap Y^{-1}(D)) = \int_{C} P_{Y|X}^{x}(D) \, dP_X(x). \tag{1}$$

As corollaries of Theorem 1 we have

Theorem 2 *Suppose that the conditional distribution of Y given X exists. For every integrable random variable $g(Y)$, let*

$$h(x) = \int g(y) \, dP_{Y|X}^{x}(y).$$

Then the random variable $h(X)$ is a determination of $E(g(Y)|X)$.

If X and Y are random vectors the conditional distribution of Y given X always exists and the following formula is very useful.

Theorem 3 *Let X and Y be two random vectors taking values on $(\mathbb{R}^s, \mathscr{B}_{\mathbb{R}^s})$ and $(\mathbb{R}^k, \mathscr{B}_{\mathbb{R}^k})$, respectively. Let ν and μ be two positive σ-finite measures defined on $(\mathbb{R}^s, \mathscr{B}_{\mathbb{R}^s})$ and $(\mathbb{R}^k, \mathscr{B}_{\mathbb{R}^k})$, respectively. If the pair (X, Y) has a probability density function f with respect to the measure $\nu \otimes \mu$, then*

$$\frac{dP_{Y|X}^{x}}{d\mu} = \frac{f(x, y)}{\displaystyle\int_{\mathbb{R}^k} f(x, y) \, d\mu}, \qquad \nu\text{-a.e.}$$

6. Transition Probabilities

The transition probability introduced below is a very useful concept in the theory of probability and statistics. For more details and applications, see Neveu [53, p. 73]. In Definition 5.2 only $P_{Y|X}^{x}$ and P_X appear; with the aid of the transition probability we shall show in this section, that they are always enough to determine $P_{(X,Y)}$.

Definition 1 Let $(\mathscr{X}, \mathscr{C})$ and $(\mathscr{Y}, \mathscr{D})$ be two probability spaces. A transition probability is a function $P^x(D)$ on $\mathscr{X} \times \mathscr{D}$ such that

(a) for every $x \in \mathscr{X}$, the function $D \to P^x(D)$ is a probability distribution on $(\mathscr{Y}, \mathscr{D})$

(b) for every $D \in \mathscr{D}$, the function $x \to P^x(D)$ is a $\mathscr{C}$-measurable function.

Theorem 1 [53, p. 74] *Let $(\mathscr{X}, \mathscr{C})$ and $(\mathscr{Y}, \mathscr{D})$ be two probability spaces and let $P^x(D)$ be a transition probability on $\mathscr{X} \times \mathscr{D}$. For every distribution Q on $(\mathscr{X}, \mathscr{C})$ there exists a unique distribution π on $(\mathscr{X} \times \mathscr{Y}, \mathscr{C} \otimes \mathscr{D})$ such that*

$$\forall C \in \mathscr{C}, \quad \forall D \in \mathscr{D}, \qquad \pi(C \times D) = \int_C P^x(D)\, dQ(x). \tag{1}$$

Then for every integrable (or positive) random variable Z on $(\mathscr{X} \times \mathscr{Y}, \mathscr{C} \otimes \mathscr{D}, \pi)$ the function

$$Z_1(x) = \int_{\mathscr{Y}} Z(x, y)\, dP^x(y) \qquad \left(\text{or } \int_{\mathscr{Y}}^{*} Z(x, y)\, dP^x(y)\right)$$

is defined Q-a.e. and is integrable (or positive and $\mathscr{C}$-measurable) with respect to Q and

$$E(Z) = \int_{\mathscr{X}} Z_1(x)\, dQ(x) \qquad \left(\text{or } \int_{\mathscr{X}}^{*} Z_1(x)\, dQ(x)\right).$$

We remark that by using notation of the previous section, one can show that (5.1) is equivalent to (1) and then $P_{Y|X}^{x}(D)$ is a transition probability on $\mathscr{X} \times \mathscr{D}$. Thus the distribution P_X and the conditional distribution $P_{Y|X}^{x}$ determine $P_{(X,Y)}$.

Theorem 2 [52, p. 89] (Fubini) *Let $(\mathscr{X}, \mathscr{C})$ and $(\mathscr{Y}, \mathscr{D})$ be two measurable spaces and let μ and μ' be two positive σ-finite measures defined on $(\mathscr{X}, \mathscr{C})$ and $(\mathscr{Y}, \mathscr{D})$, respectively. Let X be a positive measurable function on $(\mathscr{X} \times \mathscr{Y}, \mathscr{C} \otimes \mathscr{D})$. Then the functions*

$$X_1(x) = \int_{\mathscr{Y}}^{*} X(x, y)\, d\mu'(y), \qquad X_2(y) = \int_{\mathscr{X}}^{*} X(x, y)\, d\mu(x)$$

are positive and $\mathscr{C}$- and $\mathscr{D}$-measurable, respectively. Moreover,

$$\int_{\mathscr{X}\times\mathscr{Y}}^{*} X \, d(\mu \otimes \mu') = \int_{\mathscr{X}}^{*} X_1(x) \, d\mu(x) = \int_{\mathscr{Y}}^{*} X_2(y) \, d\mu'(y). \tag{2}$$

Equation (2) is to be interpreted to mean that if one of these integrals is finite, then the others are finite and equal to it.

Remark If X is assumed to be $\mu \otimes \mu'$-integrable, without necessarily being positive, then the theorem remains valid if we replace the upper integrals by ordinary integrals [52, p. 90].

References

1. Anderson, T. W., "An Introduction to Multivariate Statistical Analysis." Wiley, New York, 1958.
2. Basu, D., On statistics independent of a complete sufficient statistic. *Sankhyā* **15** (1953), 377–380; **20** (1958), 223–226.
3. Barra, J. R., Contribution à l'étude des lois de probabilités empiriques. Thesis, Paris (1961).
4. Barra, J. R., Carrés latins et eulériens. *Rev. Inst. Internat. Statist.* **33-1** (1965), 16.
5. Barra, J. R., and Baille, A., "Problèmes de Statistique Mathématique." Dunod, Paris, 1969.
6. Barra, J. R., "Notions Fondamentales de Statistique Mathématique." Dunod, Paris, 1971.
7. Beyer, R. T., "Handbook of Tables for Probability and Statistics." The Chemical Rubber Publ. Co., Cleveland, Ohio 1966.
8. Blackwell, D., and Girschick, M. A., "Theory of Games and Statistical Decisions." Wiley, New York, 1954.
9. Brown, L. D., Sufficient statistics in the case of independent random variables. *Ann. Math. Statist.* **35** (1964), 1456–1474.
10. Bourbaki, N. "Elements de Mathematique," Actualites Scientifiques et Industrielles 1189, Vol. XV, Livre 5, Chapitre II, §4, No. 2. Hermann, Paris, 1953.
11. Cochran, W. C., and Cox, D. R., "Experimental Designs." Wiley, New York, 1957.
12. Courrège, P., and Priouret, C., Sur l'évaluation des intégrales multiples exprimant le niveau des tests F en statistique multinormale, *Bull. Sci. Math., 2nd Ser., Paris.* **95** (1971).
13. Dieudonné, J., "Foundations of Modern Analysis." Academic Press, New York, 1960.
14. Dugué, D., "Traité de Statistique Théorique et Appliquée." Masson, Paris, 1958.
15. Dynkin, E. B., Necessary and sufficient statistics for a family of probability distributions. *Selected Trans. Math. Statist., Probab.* **1** (1961), 17–40.
16. Dugundji, J., "Topology." Allyn and Bacon, Rockleigh, New Jersey, 1966.

17. Darling, D. A., The Kolmogorov–Smirnov, Cramer–Von Mises tests. *Ann. Math. Statist.* **28** (1957), 823–838.
18. Ferguson, T. S. "Mathematical Statistics." Academic Press, New York, 1967.
19. Fisher, R. A., "Statistical Methods and Scientific Inference." Oliver and Boyd, London, 1956.
20. Fisher, R. A., and Yates, F., "Statistical Tables for Biological, Agricultural and Medical Research." Oliver and Boyd, London, 1963.
21. Fisher, R. A., "The Design of Experiments." Oliver and Boyd, London, 1960.
22. Fraser, R. A., "Non-parametric Methods in Statistics." Wiley, New York, 1957.
23. Feller, W., On the Kolmogorov–Smirnov limit theorems for empirical distributions, *Ann. Math. Statist.* **19** (1948), 177–189.
24. Fraser, D. A. S., "Probability and Statistics." Wadsworth (Duxbury), Belmont, California, 1976.
25. Fraser, D. A. S., "Inference and Linear Models." Mc Graw Hill, New York, 1979.
26. Good, I. J., "The estimation of probabilities. An essay on modern Bayesian methods," Research Monograph 30. MIT Press, Cambridge, Massachusetts, 1965.
27. Halmos, P. R., "Measure Theory." Van Nostrand-Reinhold, Princeton, New Jersey, 1950.
28. Hormander, L. "An Introduction to Complex Analysis in Several Variables." Van Nostrand-Reinhold, Princeton, New Jersey, 1966.
29. Hsu, P. L. Analysis of variance from the power function standpoint, *Biometrika* **32** (1941), 62–69.
30. Hanen, A., and Neveu, J., Atomes conditionnels d'un espace de probabilité. *Acta Math. Acad. Sci. Hung.* **17** (3–4) (1966), 443–449.
31. Hanani, H., The existence and construction of balanced incomplete block designs, *Ann. Math. Statist.* **32** (1961), 361.
32. Hall, M., "Combinatorial Theory." Ginn (Blaisdell), Boston, Massachusetts, 1967.
33. Hisleur, G., Une estimation optimale des paramètres d'une loi normale, *Rev. Statist. Appl.* **15** No. 3 (1967), 43.
34. Herbach, L. H., Properties of model II-type analysis of variance tests, *Ann. Math. Statist.* **30** (1959), 939–959.
35. Kagan, A. M., New classes of families of distributions allowing similar regions, *Proc. Steklov Inst. Math.* **79** (1965).
36. Kingman, J. F. C., and Robertson, A. P., On a theorem of Lyapunov, *J. London Math. Soc.* **43** (1968), 347–351.
37. Kullback, S., "Information Theory and Statistics." Wiley, New York, 1959.
38. Kendall, A. G., and Stuart, A., "The Advanced Theory of Statistics" (3 Vols.). Griffin, London, 1967, 1968, 1969.
39. Kempthorne, O., "The Design and Analysis of Experiments." Wiley, New York, 1952.
40. Linnik, Yu. V., Tests, unbiased estimates and cotest ideals, *Doklady* **161** (1965), 520; *Sov. Math. Dokl.* **6** (1965), 459.
41. Linnik, Yu. V., Statistical problems with nuisance parameters, *Trans. Amer. Math. Soc.* **20** (1968).
42. Linnik, Yu. V., "Leçons sur les Problèmes de Statistique Analytique." Gauthier-Villars, Paris, 1967.
43. Linnik, Yu. V., On the elimination of nuisance parameters in statistical problems, *Berkeley Symp. Math. Statist. Probab., 5th, 1965* **1** 267. Univ. of California Press, Berkeley, California, 1967.
44. Lehmann, E., "Testing Statistical Hypothesis." Wiley, New York, 1959.
45. Laurent, P. J., "Approximation et Optimisation." Hermann, Paris, 1972.

46. Lukacs, E., and Laha, R. G., "Applications of Characteristic Functions." Griffin, London, 1964.
47. Lojaseiewicz, S., Sur le problème de la division, *Studia Math.* **18** (1959).
48. Wilkinson, J. H., "The Algebraic Eigenvalue Problem." Oxford Univ. Press (Clarendon), London and New York, 1965.
49. Lindgren, B. W., "Statistical Theory." Macmillan, New York, 1962.
50. Massey, F. J., The Kolmogorov–Smirnov Test for Goodness of Fit, *J. Amer. Statist. Assoc.* **46** (1951), 68–78.
51. Mann, H. B., "Analysis and Design of Experiments." Dover, New York, 1949.
52. Metivier, M., "Notions Fondamentales de Théorie des Probabilités." Dunod, Paris, 1968.
53. Neveu, J., "Mathematical Foundations of the Calculus of Probability." Holden Day, San Francisco, California, 1965.
54. Narasimhan, R., "Analysis on Real and Complex Manifolds." Masson, Paris, 1968.
55. Palamodov, V. P., Testing of a multidimensional polynomial hypothesis, *Doklady* **55–172** (1967); *Sov. Math. Dokl.* **8** (1967), 95.
56. Pham-Dinh, T., Problème de test d'hypothèses linéaires multiples en analyse de la variance, *Rev. Statist. Appl.* **19,** No. 1 (1971).
57. Romanovsky, I. V., and Sudakov, V. N., On the existence of independent partitions, *Amer. Math. Soc. Proc. Steklov Inst. Math.* **79** (1965).
58. Soler, J. L., Notions de liberté en statistique mathématique. Thesis, Univ. of Grenoble, 1970.
59. Savage, L. J., "The Foundations of Statistics." Wiley, New York, 1954.
60. Scheffé, H., "The Analysis of Variance." Wiley, New York, 1959.
61. Scheffé, H., On solutions of the Behrens–Fisher problem based on the t-distribution, *Ann. Math. Statist.* **14** (1943), 35–44.
62. Tortrat, A., and Hennequin, P. L., "Théorie des Probabilités et quelques Applications." Masson, Paris, 1965.
63. Wald, A., "Statistical Decision Functions." Wiley, New York, 1950.
64. Wald, A., On the power function of the analysis of variance test, *Ann. Math. Statist.* **13** (1942), 434–439.
65. Wilks, S. S., "Mathematical Statistics." Wiley, New York, 1962.
66. Wilansky, A., "Functional Analysis." Ginn (Blaisdell), Boston, Massachusetts, 1964.

Index

Probability and Mathematical Statistics

A Series of Monographs and Textbooks

Thomas Ferguson. Mathematical Statistics: A Decision Theoretic Approach. 1967

Howard Tucker. A Graduate Course in Probability. 1967

K. R. Parthasarathy. Probability Measures on Metric Spaces. 1967

P. Révész. The Laws of Large Numbers. 1968

H. P. McKean, Jr. Stochastic Integrals. 1969

B. V. Gnedenko, Yu. K. Belyayev, and A. D. Solovyev. Mathematical Methods of Reliability Theory. 1969

Demetrios A. Kappos. Probability Algebras and Stochastic Spaces. 1969

Ivan N. Pesin. Classical and Modern Integration Theories. 1970

S. Vajda. Probabilistic Programming. 1972

Sheldon M. Ross. Introduction to Probability Models. 1972

Robert B. Ash. Real Analysis and Probability. 1972

V. V. Fedorov. Theory of Optimal Experiments. 1972

K. V. Mardia. Statistics of Directional Data. 1972

H. Dym and H. P. McKean. Fourier Series and Integrals. 1972

Tatsuo Kawata. Fourier Analysis in Probability Theory. 1972

Fritz Oberhettinger. Fourier Transforms of Distributions and Their Inverses: A Collection of Tables. 1973

Paul Erdös and Joel Spencer. Probabilistic Methods in Combinatorics. 1973

K. Sarkadi and I. Vincze. Mathematical Methods of Statistical Quality Control. 1973

Michael R. Anderberg. Cluster Analysis for Applications. 1973

W. Hengartner and R. Theodorescu. Concentration Functions. 1973

Kai Lai Chung. A Course in Probability Theory, Second Edition. 1974

L. H. Koopmans. The Spectral Analysis of Time Series. 1974

L. E. Maistrov. Probability Theory: A Historical Sketch. 1974

William F. Stout. Almost Sure Convergence. 1974

E. J. McShane. Stochastic Calculus and Stochastic Models. 1974

Robert B. Ash and Melvin F. Gardner. Topics in Stochastic Processes. 1975

Avner Friedman, Stochastic Differential Equations and Applications, Volume 1, 1975; Volume 2. 1975

Roger Cuppens. Decomposition of Multivariate Probabilities. 1975

Eugene Lukacs. Stochastic Convergence, Second Edition. 1975

H. Dym and H. P. McKean. Gaussian Processes, Function Theory, and the Inverse Spectral Problem. 1976

N. C. Giri. Multivariate Statistical Inference. 1977

Lloyd Fisher and John McDonald. Fixed Effects Analysis of Variance. 1978

Sidney C. Port and Charles J. Stone. Brownian Motion and Classical Potential Theory. 1978

Konrad Jacobs. Measure and Integral. 1978

K. V. Mardia, J. T. Kent, and J. M. Biddy. Multivariate Analysis. 1979

Sri Gopal Mohanty. Lattice Path Counting and Applications. 1979

Y. L. Tong. Probability Inequalities in Multivariate Distributions. 1980

Michel Metivier and J. Pellaumail. Stochastic Integration. 1980

M. B. Priestly, Spectral Analysis and Time Series. 1980

Ishwar V. Basawa and B. L. S. Prakasa Rao, Statistical Inference for Stochastic Processes. 1980

M. Csörgö and P. Révész. Strong Approximations in Probability and Statistics. 1980

Sheldon Ross. Introduction to Probability Models, Second Edition. 1980

P. Hall and C. C. Heyde. Martingale Limit Theory and Its Application. 1980

Imre Csiszár and János Körner, Information Theory: Coding Theorems for Discrete Memoryless Systems. 1981

A. Hald. Statistical Theory of Sampling Inspection by Attributes. 1981

H. Bauer. Probability Theory and Elements of Measure Theory. 1981

M. M. Rao. Foundations of Stochastic Analysis. 1981

Jean-Rene Barra. Mathematical Basis of Statistics. Translated and Edited by L. Herbach. 1981